T0392972

Materials Science to Combat COVID-19

This book covers the application of emerging materials to combat coronavirus. It discusses various physicochemical and biomedical characteristics of 2D materials, such as graphene, MXenes, and their various other derivatives, followed by proposal of how such materials can be used to design and develop several systems capable of fighting infectious diseases like coronavirus. It also covers fundamental and product developments based on MXene and graphene-based materials using emerging 3D printing process and other pertinent aspects.

Features:

- Focusses on COVID-19 from cross-disciplinary approach, especially biophysical management of the virus.
- Discusses emerging 2D materials such as MXene and graphene to combat coronavirus.
- Reviews development of PPEs, sterilizers, foams, antimicrobial surfaces, biosensors from materials science perspective.
- Explores pertinent fundamental mechanisms to develop structure–property relationships.
- Examines cytotoxicity and biocompatibility of the discussed 2D materials.

This book is aimed at researchers and graduate students in biotechnology, biomedical engineering, chemical engineering, and materials science.

Emerging Materials and Technologies

Series Editor: Boris I. Kharissov

The *Emerging Materials and Technologies* series is devoted to highlighting publications centered on emerging advanced materials and novel technologies. Attention is paid to those newly discovered or applied materials with potential to solve pressing societal problems and improve quality of life, corresponding to environmental protection, medicine, communications, energy, transportation, advanced manufacturing, and related areas.

The series takes into account that, under present strong demands for energy, material, and cost savings, as well as heavy contamination problems and worldwide pandemic conditions, the area of emerging materials and related scalable technologies is a highly interdisciplinary field, with the need for researchers, professionals, and academics across the spectrum of engineering and technological disciplines. The main objective of this book series is to attract more attention to these materials and technologies and invite conversation among the international R&D community.

Smart Nanomaterials
Imalka Munaweera and M. L. Chamalki Madhusha

Nanocosmetics
Drug Delivery Approaches, Applications and Regulatory Aspects
Edited by: Prashant Kesharwani and Sunil Kumar Dubey

Sustainability of Green and Eco-friendly Composites
Edited by Sumit Gupta, Vijay Chaudhary and Pallav Gupta

Assessment of Polymeric Materials for Biomedical Applications
Edited by Vijay Chaudhary, Sumit Gupta, Pallav Gupta and Partha Pratim Das

Nanomaterials for Sustainable Energy Applications
Edited by Piyush Kumar Sonkar and Vellaichamy Ganesan

Materials Science to Combat COVID-19
Edited by Neeraj Dwivedi and Avanish Kumar Srivastava

Two-Dimensional Nanomaterials for Fire-Safe Polymers
Yuan Hu and Xin Wang

For more information about this series, please visit: www.routledge.com/Emerging-Materials-and-Technologies/book-series/CRCEMT

Materials Science to Combat COVID-19

Edited by Neeraj Dwivedi and
Avanish Kumar Srivastava

CRC Press
Taylor & Francis Group
Boca Raton London New York

CRC Press is an imprint of the
Taylor & Francis Group, an **informa** business

First edition published 2024
by CRC Press
6000 Broken Sound Parkway NW, Suite 300, Boca Raton, FL 33487–2742

and by CRC Press
4 Park Square, Milton Park, Abingdon, Oxon, OX14 4RN

CRC Press is an imprint of Taylor & Francis Group, LLC
© 2024 selection and editorial matter, Neeraj Dwivedi and Avanish Kumar Srivastava; individual chapters, the contributors

ISBN: 9781032327204 (hbk)
ISBN: 9781032327211 (pbk)
ISBN: 9781003316381 (ebk)

DOI: 10.1201/9781003316381

Typeset in Times
by Apex CoVantage, LLC

Contents

About the Editors

Neeraj Dwivedi is currently a principal scientist at CSIR – Advanced Materials and Processes Research Institute (CSIR-AMPRI), Bhopal, India (since June 2019), and associate professor at AcSIR, India. He completed his PhD from Indian Institute of Technology (IIT) Delhi, India, in 2013 and then moved to National University of Singapore, where he worked for about six years (2013–2019) as a postdoctoral fellow. Dr. Dwivedi works on 2D materials, thin films, and polymer composites for tribology, shape memory, biomedical, electronic, and energy applications. He has published one book chapter, about 95 research papers in highly reputed international journals, many book chapters, and more than 100 research papers; served as a lead guest editor for one of the special issues of *Journal of Nanomaterials* (Hindawi, 2015); and is currently a reviewer of more than 20 international journals. As per Google Scholar, he has a total number of 3,240 citations; h-index 31; and i-10 index 73.

Avanish Kumar Srivastava is currently the director of CSIR – Advanced Materials and Processes Research Institute (CSIR-AMPRI), Bhopal, India. Presently, he is also the president of electron microscopy society of India. He completed his PhD from Indian Institute of Science (IISc) in 1996 and then joined the National Physical Laboratory (NPL), India, as a scientist. His current research interest includes characterization of nanomaterials using electron microscopy and exploration of nanomaterials, graphene-based materials, and metal oxides for wide-spectrum applications. Prof. Srivastava has published about 300 research articles in reputed international journals and has been invited for talks in different scientific meetings in India and abroad; 12 patents have been filed/granted into his credit, along with 11 technology/know-how transfers to various private partners. He is a member/fellow of several research and professional societies of India and abroad.

Contributors

Jamana Prasad Chaurasia
CSIR – Advanced Materials and
 Processes Research Institute
Bhopal, India

Chetna Dhand
CSIR – Advanced Materials and
 Processes Research Institute
Bhopal, India

Neeraj Dwivedi
CSIR – Advanced Materials and
 Processes Research Institute
Bhopal, India

Gaurav Kumar Gupta
CSIR – Advanced Materials and
 Processes Research Institute
Bhopal, India

Manoj Kumar Gupta
CSIR – Advanced Materials and
 Processes Research Institute
Bhopal, India

Pradip Kumar
CSIR – Advanced Materials and
 Processes Research Institute
Bhopal, India

Rajeev Kumar
CSIR – Advanced Materials and
 Processes Research Institute
Bhopal, India

Avanish Kumar Srivastava
CSIR – Advanced Materials and
 Processes Research Institute
Bhopal, India

1 Introduction

Neeraj Dwivedi and Avanish Kumar Srivastava

In December 2019, an outburst of pneumonia began due to a new strain of coronavirus which the World Health Organization (WHO) named the 2019 novel coronavirus (2019-nCoV) [1–3]. The virus was found to be highly infectious, and within 3–4 months of time period, it covered almost the entire world. In a formal meeting on February 11, 2022, WHO renamed the virus "severe acute respiratory syndrome coronavirus 2 (SARS-CoV-2)" because the structure of SARS-CoV-2 was examined to be similar to the previously discovered SARS coronavirus [4]. The disease as a consequence of SARS-CoV-2 was named the coronavirus disease-19 (COVID-19) [4]. Given its significant adverse effects on human health globally, WHO professed COVID-19 a pandemic. The common symptoms of COVID-19 were noted to be fever, cold, cough, throat pain, headache, pneumonia, body pain, etc. [1, 2].

COVID-19 is not the first pandemic that the world has witnessed ever. Humans have faced endemics or pandemics almost in every century. One of the earliest endemics/pandemics, encountered in 430 BC in Athens, which is one of the major cities of Greece, was the Plague of Athens [5, 6]. The disease was observed to be so fatal that it led to the death of about 25% of the population of the city. Some of the major endemics or pandemics encountered in regular intervals are listed in Table 1.1.

Materials science, which is a cross-disciplinary field of science and engineering that deals with a wide variety of materials for numerous applications, has definitely played a vital role in combating COVID-19 [7–9]. Conventional and novel materials, such as graphene and graphene-related materials (GRMs) [9–11], MXene [8, 12], metal/metal oxides [13], metal nitrides [8, 14, 15], polymers [16–18], intelligent materials [19, 20], and so on, have been employed for the development of diagnostic, treatment, preventive, and safety components/systems for tackling COVID-19.

In particular, 2D materials, such as graphene and GRMs, that show exceptional physicochemical and functional properties, including antimicrobial characteristic, high mobility and conductivity, large surface area, and outstanding piezoelectric, mechanical, electrochemical, photocatalytic, photothermal, sensing, and electromagnetic shielding properties, are at the center of new technology development [21–25]. These materials also offer functionalization feasibility and integration with emerging manufacturing technologies such as 3D printing [10]. Thus, due to the excellent characteristics of graphene and GRMs, recently, many review, perspective, and original articles have appeared in the literature that show how these materials can contribute to combating COVID-19. Palmeri and Papi [9] discussed about the possible interaction of graphene and GRMs with viruses; it was suggested that graphene attached with antibody can promptly detect the virus protein and hence can be exploited for development of biosensor for the fast screening of infected people at the mass level. They also suggested that graphene and GRMs can be exploited to develop filters, sterilizers

DOI: 10.1201/9781003316381-1

TABLE 1.1
List of some of the major endemics/pandemics that the world has seen so far

Endemic or Pandemic	Year/Remark
Plague of Athens	430 BC. One of the earliest known endemics/pandemics.
Antonine Plague	AD 165–180. This is also known as the Plague of Galen, which impacted the Roman Empire.
Plague of Cyprian	AD 249–262. This also affected the Roman Empire.
Leprosy	It is also known as Hansen's disease after the discovery of causative bacteria of leprosy by G. H. Armauer Hansen. Leprosy is caused by the *Mycobacterium leprae* or *Mycobacterium lepromatosis* bacteria.
The black death	1346–1353. This was one of the most fatal pandemics in human history that killed about 75–200 million population in North Africa and Eurasia.
Columbian exchange	Columbian exchange diseases initially appeared around 1490–1495 with the spread of syphilis from the people of the Caribbean Sea to Europe.
The Great Plague of London	1665–1666. The Great Plague of London was caused by *Yersinia pestis* bacteria. This largely affected the population of London, with death toll exceeding 65–70 thousand.
First cholera pandemic	1817–1824. The first cholera pandemic, also called the first Asiatic cholera pandemic, started in West Bengal, India, and spread across Southeast Asia, the Middle East, Eastern Africa, etc. It was a very fatal disease at that time, and it led to a massive number of deaths in affected countries.
The third plague pandemic	1855. It was a bubonic plague pandemic that started in China. It was a lethal pandemic that caused the deaths of over 15 million people worldwide, with huge loss of human lives in India and China.
Fiji measles	1875. It was a highly contagious infectious disease caused by the measles virus. The disease spread across Fiji in 1875 with an exceptional pace. It killed several tens of thousands of people in Fiji alone.
Russian flu	1889–1890. Russian flu is also called the 1889–1890 pandemic or Asiatic flu. It was one of the deadliest viral pandemics that caused the death of about 1.5 billion people globally.
Spanish flu	1918. It is also referred to as the 1918 influenza pandemic caused by H1N1 influenza virus. It started in 1918 and was considered one of the highly deadliest pandemics. The death tolls due to Spanish flu reached about 50–100 million globally.
Asian flu	1957. It was a flu pandemic caused by the avian influenza virus.
The human immunodeficiency virus infection/acquired immune deficiency syndrome (HIV/AIDS)	1981. AIDS is a viral disease caused by HIV. Among the two strains of HIV, namely, HIV-1 and HIV-2, HIV-1 is highly transmissible and responsible for most of the AIDS cases in the world.

TABLE 4.1 (CONTINUED)

Endemic or Pandemic	Year/Remark
Severe acute respiratory syndrome (SARS)	2003. It is a viral respiratory disease caused by SARS-coronavirus (SARS-CoV).
COVID-19	2019. It is also a viral respiratory disease caused by a new strain of coronavirus called SARS-CoV-2. The disease due to SARS-CoV is referred to as COVID-19. Given its huge adverse effects on human health, exceptional works have been performed for vaccine development and materials management to combat the COVID-19 pandemic. Finally, scientists have been able to develop vaccines for COVID-19 in a very short period of time, and at the same time, various technologies have evolved based on advanced materials and processes to combat COVID-19.

Note: The details of some of these endemics/pandemics can be found in ref. [5,6], Wikipedia, and various other online documents.

(due to their photocatalytic and photothermal characteristics), and many other personal protective equipment (PPEs). Cordaro et al. [26] critically discussed the role of graphene and GRMs in the diagnosis of viral diseases and proposed how these materials can be exploited to develop SARS-CoV-2 sensor. Graphene-based sensors have been developed for the detection of SARS-CoV-2 [27, 28]. Moreover, interaction of SARS-CoV-2 with graphene oxide has led to the killing/inactivation of the virus [29].

Furthermore, MXene-based materials are also an emerging class of materials which display astonishing physicochemical and functional properties, including antiviral and antibacterial properties, high electrical conductivity, and excellent photothermal, photocatalytic, and mechanical properties [12, 30–33]. Moreover, surface plasmon resonance effect has also been observed in graphene, GRMs, and MXene-based materials [34–38]. Further, due to synergistic effects, the hybrid of graphene–MXene and GRMs–MXene can be promising candidates for electronic, optical, mechanical, biomedical, sensor, shielding, and energy applications [12, 39–42]. Additionally, the inclusion of graphene, GRMs, and MXene-based materials as reinforcement in the metal, ceramic, and polymer matrix can lead to enhancement of several physicochemical and functional characteristics of host materials [12, 43–45]. These materials, therefore, can be explored to develop various technologies for fighting against COVID-19. Moreover, MXene or graphene–MXene hybrid materials have been recently developed for the fast detection of SARS-CoV-2 [46].

Transition metal dichalcogenides (TMDCs) are also emerging 2D materials for a range of applications. Unlike graphene, TMDCs, for example, MoS_2, possess bandgap, display semiconducting behavior, and have good electrical, thermal, mechanical, photothermal, photocatalytic, and various other functional properties [47–50]. Wei et al. developed an MoS_2-based field effect transistor (FET) for the detection of SARS-CoV-2 antibody [51]. The developed device was found to be highly sensitive and specific for SARS-CoV-2 detection. Kumar et al. developed an MoS_2-engineered reusable antimicrobial fabric for PPE development [52]. They observed that MoS_2-modified fabric maintained antimicrobial characteristic even after 60 washing cycles. Thus, in general, due to exceptional properties, graphene; GRMs; TMDCs, such as MoS_2, MXene, graphene/GRMs---MXene hybrids; and composites of these materials, have huge potential for biomedical technologies, including controlling the current and future endemic/pandemic. These materials can be used to develop preventive/protective components (such as filters, face mask, aprons, face shield, etc.), diagnostic components (such as electrochemical, piezoelectric, and field effect transistor (FET) based biosensors for SARS-CoV-2 and other virus detection), and treatment components (such as antiviral coatings on high-touch surfaces, antiviral filters, antiviral cloths, sterilizers, and so on for early inactivation/killing of the viruses).

This book focuses on materials science approach, in particular 2D materials, to combat COVID-19. Given the subject COVID-19, it is of paramount interest to know and understand the coronavirus family in detail. Therefore, Chapter 2 is about the coronavirus family, their biological structures, and how they interact with humans. Chapter 3 describes the materials synthesis and properties of major 2D materials, including graphene, GO, rGO, TMDCs, MXene, MAX phases, etc. Both the physical and chemical methods are discussed for their synthesis. Additionally, the fundamental and numerous functional properties of these 2D materials are also discussed in a comprehensive manner. Further, materials with antimicrobial characteristic are crucial for controlling viral spread. Therefore, Chapter 4 comprehensively describes the basics of microbial contamination and antimicrobial properties of proposed 2D materials and their composites in order to combat viral diseases, including COVID-19. The fundamental mechanisms responsible for antimicrobial activity of these 2D materials are also included in this chapter. Further, biosensors have shown great promise for early-stage diagnosis of several infections. Therefore, Chapter 5 presents the discussion about advanced biosensors based on 2D materials for early diagnostic of coronavirus to control COVID-19. The fundamentals of various types of biosensors, namely, electrochemical and FET- and piezoelectric-based biosensors developed using emerging materials, namely, graphene, GRMs, MXene, and their hybrids, and composites for the early-stage diagnosis of infections, including viral infections, are discussed. How these 2D materials–based biosensors can be better than conventional methods for the detection of coronavirus is also elaborated in this chapter. Additive manufacturing or 3D printing has emerged as a promising manufacturing technology for the facile development of complex structures/components. This method has a huge potential in the biomedical field too, in particular, in the manufacturing of implants, PPEs, and various other biomedical components. Therefore, Chapter 6 discusses about the 3D printing method to design and develop components for COVID-19. In particular, the fundamentals of additive manufacturing, its applicability for emerging materials, and the possible component/system development to combat coronavirus are discussed. Specifically, the development of PPEs and other components based on 2D materials by additive manufacturing is presented. Next, foam-type materials are an important class of materials, and due to their light weight, filtration, catalytic, and various other functional characteristics, they can be exploited for the development of healthcare products. So, the Chapter 7 presents a comprehensive overview of a variety of foams, how they can be manufactured, and their functional properties. Subsequently, how 2D materials–based foams and related various components/systems can be designed for COVID-19 and other healthcare applications is also discussed. Further, the proposed 2D materials display excellent light-to-heat conversion and photocatalytic properties which can also be harnessed to develop various healthcare systems, including sterilizers, that can contribute to minimizing COVID-19 spread. Hence, in a perspective, in Chapter 8, firstly, the photo-to-heat conversion and photocatalytic properties of graphene, GRMs, and MXene-based materials are discussed, and then how they can be used to develop high-performance sterilizers is presented. In order to explore the materials for biomedical applications, the important parameters include biocompatibility and cytotoxicity. Thus, Chapter 9 explores the biocompatibility and cytotoxicity of emerging 2D materials, such as graphene, MXene, and related materials. Finally, in Chapter 10, we conclude the characteristics of emerging materials in the context of biomedical applications and discuss future prospects. Figure 1.1 illustrates the major characteristics and applications of 2D materials discussed in the book for combating COVID-19.

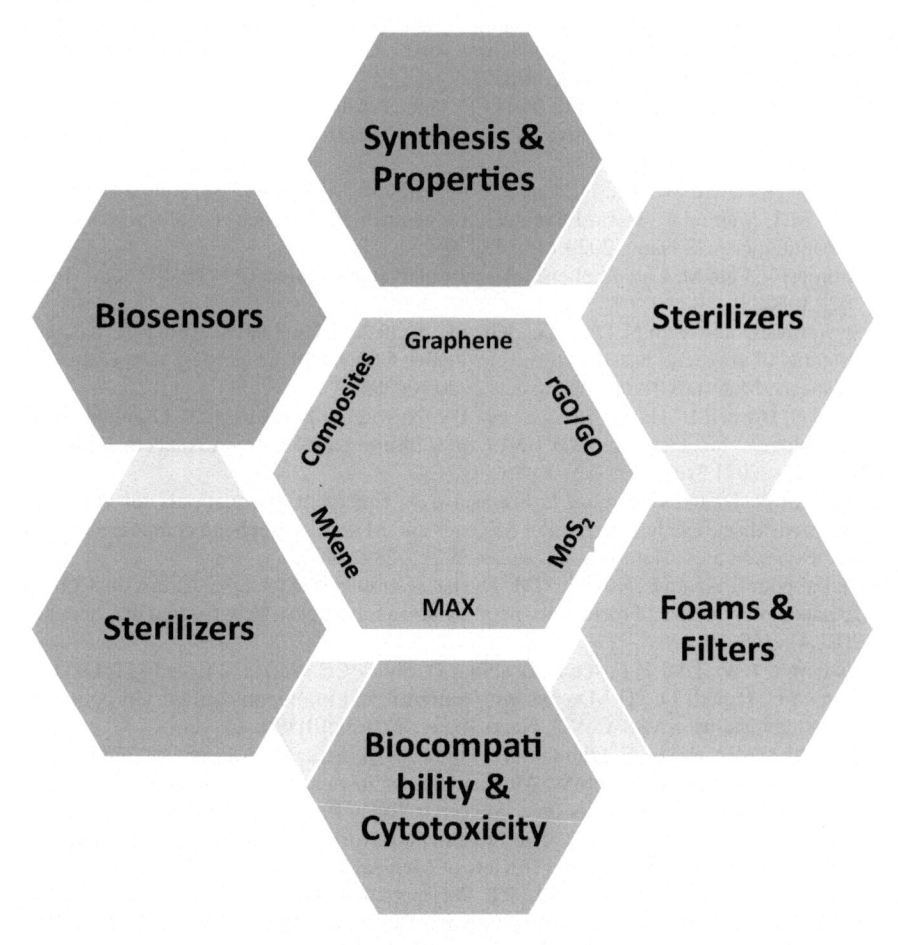

FIGURE 1.1 Schematic illustration for synthesis, properties, and different applications of diverse 2D materials and their composites which will be discussed in this book.

REFERENCES

1. Fernandes Q, Inchakalody VP, Merhi M, Mestiri S, Taib N, Moustafa Abo El-Ella D, Bedhiafi T, Raza A, Al-Zaidan L, Mohsen MO. Emerging COVID-19 variants and their impact on SARS-CoV-2 diagnosis, therapeutics and vaccines. Annals of Medicine. 2022;54:524–40.
2. Walls AC, Park Y-J, Tortorici MA, Wall A, McGuire AT, Veesler D. Structure, function, and antigenicity of the SARS-CoV-2 spike glycoprotein. Cell. 2020;181:281–92.e6.
3. Shang J, Ye G, Shi K, Wan Y, Luo C, Aihara H, Geng Q, Auerbach A, Li F. Structural basis of receptor recognition by SARS-CoV-2. Nature. 2020;581:221–4.
4. Dwivedi N, Yeo RJ, Zhang Z, Dhand C, Tripathy S, Bhatia CS. Interface engineering and controlling the friction and wear of ultrathin carbon films: high sp3 versus high sp2 carbons. Advanced Functional Materials. 2016;26:1526–42.
5. Dwivedi N, Yeo RJ, Zhang Z, Dhand C, Tripathy S, Bhatia CS. Direct observation of thickness and foreign interlayer driven abrupt structural transformation in ultrathin carbon and hybrid silicon nitride/carbon films. Carbon. 2017;115:701–19.
6. Dwivedi N, Yeo RJ, Yak LJ, Satyanarayana N, Dhand C, Bhat TN, Zhang Z, Tripathy S, Bhatia CS. Atomic scale interface manipulation, structural engineering, and their impact on ultrathin carbon films in controlling wear, friction, and corrosion. ACS Applied Materials & Interfaces. 2016;8:17606–21.
7. Tang Z, Kong N, Zhang X, Liu Y, Hu P, Mou S, Liljeström P, Shi J, Tan W, Kim JS. A materials-science perspective on tackling COVID-19. Nature Reviews Materials. 2020;5:847–60.
8. Weiss C, Carriere M, Fusco L, Capua I, Regla-Nava JA, Pasquali M, Scott JA, Vitale F, Unal MA, Mattevi C. Toward nanotechnology-enabled approaches against the COVID-19 pandemic. ACS Nano. 2020;14:6383–406.
9. Palmieri V, Papi M. Can graphene take part in the fight against COVID-19? Nano Today. 2020:100883.
10. Srivastava A, Dwivedi N, Dhand C, Khan R, Sathish N, Gupta MK, Kumar R, Kumar S. Potential of graphene-based materials to combat COVID-19: properties, perspectives and prospects. Materials Today Chemistry. 2020:100385.
11. Afroj S, Britnell L, Hasan T, Andreeva DV, Novoselov KS, Karim N. Graphene-based technologies for tackling COVID-19 and future pandemics. Advanced Functional Materials. 2021:2107407.
12. Dwivedi N, Dhand C, Kumar P, Srivastava A. Emergent 2D materials for combating infectious diseases: the potential of MXenes and MXene—graphene composites to fight against pandemics. Materials Advances. 2021;2:2892–905.
13. Mallakpour S, Azadi E, Hussain CM. The latest strategies in the fight against the COVID-19 pandemic: the role of metal and metal oxide nanoparticles. New Journal of Chemistry. 2021;45:6167–79.
14. Unal MA, Bayrakdar F, Fusco L, Besbinar O, Shuck CE, Yalcin S, Erken MT, Ozkul A, Gurcan C, Panatli O. 2D MXenes with antiviral and immunomodulatory properties: a pilot study against SARS-CoV-2. Nano Today. 2021;38:101136.
15. Pezzotti G, Boschetto F, Ohgitani E, Fujita Y, Shin-Ya M, Adachi T, Yamamoto T, Kanamura N, Marin E, Zhu W. Mechanisms of instantaneous inactivation of SARS-CoV-2 by silicon nitride bioceramic. Materials Today Bio. 2021;12:100144.
16. Zuniga JM, Cortes A. The role of additive manufacturing and antimicrobial polymers in the COVID-19 pandemic. Expert Review of Medical Devices. 2020;17:477–81.
17. Gadhave RV, Vineeth S, Gadekar PT. Polymers and polymeric materials in covid-19 pandemic: a review. Open Journal of Polymer Chemistry. 2020;10:66.
18. Mahat MM, Sabere ASM, Azizi J, Amdan NAN. Potential applications of conducting polymers to reduce secondary bacterial infections among COVID-19 patients: a review. Emergent Materials. 2021;4:279–92.

19. Yalcin HC, Kaushik A. Support of intelligent emergent materials to combat COVID-19 pandemic. Emergent Materials. 2021;4:1–2.
20. Erdem Ö, Derin E, Sagdic K, Yilmaz EG, Inci F. Smart materials-integrated sensor technologies for COVID-19 diagnosis. Emergent Materials. 2021;4:169–85.
21. Geim A, Novoselov K. The rise of graphene. Nature Materials. 2007;6:183–91.
22. Zhang B, Wang Y, Liu J, Zhai G. Recent developments of phototherapy based on graphene family nanomaterials. Current Medicinal Chemistry. 2017;24:268–91.
23. Huang L, Xu S, Wang Z, Xue K, Su J, Song Y, Chen S, Zhu C, Tang BZ, Ye R. Self-reporting and photothermally enhanced rapid bacterial killing on a laser-induced graphene mask. ACS Nano. 2020;14:12045–53.
24. Shahzad F, Kumar P, Yu S, Lee S, Kim Y-H, Hong SM, Koo CM. Sulfur-doped graphene laminates for EMI shielding applications. Journal of Materials Chemistry C. 2015;3:9802–10.
25. Ferrari AC, Bonaccorso F, Fal'Ko V, Novoselov KS, Roche S, Bøggild P, Borini S, Koppens FH, Palermo V, Pugno N. Science and technology roadmap for graphene, related two-dimensional crystals, and hybrid systems. Nanoscale. 2015;7:4598–810.
26. Cordaro A, Neri G, Sciortino MT, Scala A, Piperno A. Graphene-based strategies in liquid biopsy and in viral diseases diagnosis. Nanomaterials. 2020;10:1014.
27. Park I, Lim J, You S, Hwang MT, Kwon J, Koprowski K, Kim S, Heredia J, Stewart de Ramirez SA, Valera E. Detection of SARS-CoV-2 virus amplification using a crumpled graphene field-effect transistor biosensor. ACS Sensors. 2021;6:4461–70.
28. Seo G, Lee G, Kim MJ, Baek S-H, Choi M, Ku KB, Lee C-S, Jun S, Park D, Kim HG. Rapid detection of COVID-19 causative virus (SARS-CoV-2) in human naso-pharyngeal swab specimens using field-effect transistor-based biosensor. ACS Nano. 2020;14:5135–42.
29. Fukuda M, Islam MS, Shimizu R, Nassar H, Rabin NN, Takahashi Y, Sekine Y, Lindoy LF, Fukuda T, Ikeda T. Lethal interactions of SARS-CoV-2 with graphene oxide: implications for COVID-19 treatment. ACS Applied Nano Materials. 2021;4:11881–7.
30. Gogotsi Y, Anasori B. The rise of MXenes. ACS Nano. 2019;13:8491–4.
31. Rasool K, Helal M, Ali A, Ren CE, Gogotsi Y, Mahmoud KA. Antibacterial activity of Ti3C2T x MXene. ACS Nano. 2016;10:3674–84.
32. Choi G, Shahzad F, Bahk YM, Jhon YM, Park H, Alhabeb M, Anasori B, Kim DS, Koo CM, Gogotsi Y. Enhanced terahertz shielding of MXenes with nano-metamaterials. Advanced Optical Materials. 2018;6:1701076.
33. Naguib M, Mochalin VN, Barsoum MW, Gogotsi Y. 25th anniversary article: MXenes: a new family of two-dimensional materials. Advanced Materials. 2014;26:992–1005.
34. Salihoglu O, Balci S, Kocabas C. Plasmon-polaritons on graphene-metal surface and their use in biosensors. Applied Physics Letters. 2012;100:213110.
35. Zhang H, Sun Y, Gao S, Zhang J, Zhang H, Song D. A novel graphene oxide-based surface plasmon resonance biosensor for immunoassay. Small. 2013;9:2537–40.
36. Wu L, You Q, Shan Y, Gan S, Zhao Y, Dai X, Xiang Y. Few-layer Ti3C2Tx MXene: a promising surface plasmon resonance biosensing material to enhance the sensitivity. Sensors and Actuators B: Chemical. 2018;277:210–5.
37. Chen R, Kan L, Duan F, He L, Wang M, Cui J, Zhang Z, Zhang Z. Surface plasmon resonance aptasensor based on niobium carbide MXene quantum dots for nucleocapsid of SARS-CoV-2 detection. Microchimica Acta. 2021;188:1–10.
38. Aïssa B, Ali A, Mahmoud K, Haddad T, Nedil M. Transport properties of a highly conductive 2D Ti3C2Tx MXene/graphene composite. Applied Physics Letters. 2016;109:043109.
39. Zhao S, Zhang H-B, Luo J-Q, Wang Q-W, Xu B, Hong S, Yu Z-Z. Highly electrically conductive three-dimensional Ti3C2T x MXene/reduced graphene oxide hybrid aerogels with excellent electromagnetic interference shielding performances. ACS Nano. 2018;12:11193–202.

40. Liu Y, Yu J, Guo D, Li Z, Su Y. Ti3C2Tx MXene/graphene nanocomposites: synthesis and application in electrochemical energy storage. Journal of Alloys and Compounds. 2020;815:152403.
41. Yang Y, Cao Z, He P, Shi L, Ding G, Wang R, Sun J. Ti3C2Tx MXene-graphene composite films for wearable strain sensors featured with high sensitivity and large range of linear response. Nano Energy. 2019;66:104134.
42. Ma W, Chen H, Hou S, Huang Z, Huang Y, Xu S, Fan F, Chen Y. Compressible highly stable 3D porous MXene/GO foam with a tunable high-performance stealth property in the terahertz band. ACS Applied Materials & Interfaces. 2019;11:25369–77.
43. Zheng K, Li S, Jing L, Chen PY, Xie J. Synergistic antimicrobial titanium carbide (MXene) conjugated with gold nanoclusters. Advanced Healthcare Materials. 2020;9:2001007.
44. Yoon J, Shin M, Lim J, Lee J-Y, Choi J-W. Recent advances in MXene nanocomposite-based biosensors. Biosensors. 2020;10:185.
45. Tan K, Samylingam L, Aslfattahi N, Saidur R, Kadirgama K. Optical and conductivity studies of polyvinyl alcohol-MXene (PVA-MXene) nanocomposite thin films for electronic applications. Optics & Laser Technology. 2021;136:106772.
46. Li Y, Peng Z, Holl NJ, Hassan MR, Pappas JM, Wei C, Izadi OH, Wang Y, Dong X, Wang C. MXene—graphene field-effect transistor sensing of influenza virus and SARS-CoV-2. ACS Omega. 2021;6:6643–53.
47. Chhowalla M, Shin HS, Eda G, Li L-J, Loh KP, Zhang H. The chemistry of two-dimensional layered transition metal dichalcogenide nanosheets. Nature Chemistry. 2013;5:263–75.
48. Wang H, Li C, Fang P, Zhang Z, Zhang JZ. Synthesis, properties, and optoelectronic applications of two-dimensional MoS2 and MoS2-based heterostructures. Chemical Society Reviews. 2018;47:6101–27.
49. Li X, Zhu H. Two-dimensional MoS2: properties, preparation, and applications. Journal of Materiomics. 2015;1:33–44.
50. Manzeli S, Ovchinnikov D, Pasquier D, Yazyev OV, Kis A. 2D transition metal dichalcogenides. Nature Reviews Materials. 2017;2:17033.
51. Wei J, Zhao Z, Luo F, Lan K, Chen R, Qin G. Sensitive and quantitative detection of SARS-CoV-2 antibodies from vaccinated serum by MoS_2-field effect transistor. 2D Materials. 2021;9:015030.
52. Kumar P, Roy S, Sarkar A, Jaiswal A. Reusable MoS2-modified antibacterial fabrics with photothermal disinfection properties for repurposing of personal protective masks. ACS Applied Materials & Interfaces. 2021;13:12912–27.

2 Coronavirus and COVID-19

Chetna Dhand

In December 2019, WHO received notification of a sudden pneumonia outbreak in Wuhan, China, and on January 30, 2020, WHO declared the extremely infectious severe acute respiratory syndrome coronavirus 2 (SARS-CoV-2) virus outbreak a public health emergency of international concern (PHEIC) [1]. On February 11, 2020, the WHO designated this pandemic as coronavirus disease-2019 (COVID-19) [2], and the International Committee on Virus Taxonomy (ICTV) named the virus as SARS-CoV-2 [3]. This pandemic posed an unanticipated public health and economic threat to people all across the world. COVID-19 had caused 0.48 billion confirmed cases and 6.1 million fatalities as of April 1, 2022 [4]. Figure 2.1 shows the timeline for various historic events related to COVID-19.

Coronaviruses belong to the family *Coronaviridae*, with *Coronavirinae* subfamily. *Alphacoronavirus, Betacoronavirus, Gammacoronavirus*, and *Deltacoronavirus* are the four genera that constitute the coronavirus family [5]. Six HCoV species were known to induce infection in humans prior to December 2019 with four of them causing modest flu-like symptoms in immunocompromised people: HcoV-OC43 and HcoV-229E (detected in 1960) [6], HcoV-HKU1 (identified in 2005) [7], HCoV-NL63 (diagnosed in 2004) [8], SARS-CoV (diagnosed in 2003) [9], and MERS-CoV (discovered in 2012) [10]. The latter two coronavirus strains caused massive mortalities. Following SARS-CoV and MERS-CoV, SARS-CoV-2 is the third pathogenic human coronavirus to evolve this century [11]. According to a 2019 phylogenetic study of SARS-CoV-2 and other coronaviruses, SARS-CoV-2 shares 96%, 79.5%, and 55% similarity with the bat coronavirus RaTG13, SARS-CoV BJ01, and MERS-CoV HCoV-EMC, respectively, and therefore pertains to the same virus family that caused SARS and MERS. Both infectious illnesses have generated over 10,000 cases and have substantial fatality rates, with SARS having a 10% mortality rate and MERS having a 37% mortality rate.[12] This suggests that bats could be an origin of SARS-CoV-2, which could be transmitted to people directly or through an unknown intermediary host. Figure 2.1 presents the history of major historical events for SARS-CoV-2.

2.1 CORONAVIRUS FAMILY

Coronaviruses are huge, spherical, and positive-sense enveloped viruses which infect a variety of mammals and birds [12]. The coronavirus genome is constituted of ~3,000 nucleotides with size ranging from 26 to 32 kilobases [13]. Coronaviruses get their name from the glycoproteins-based spike like projections lying on the virus surface

DOI: 10.1201/9781003316381-2

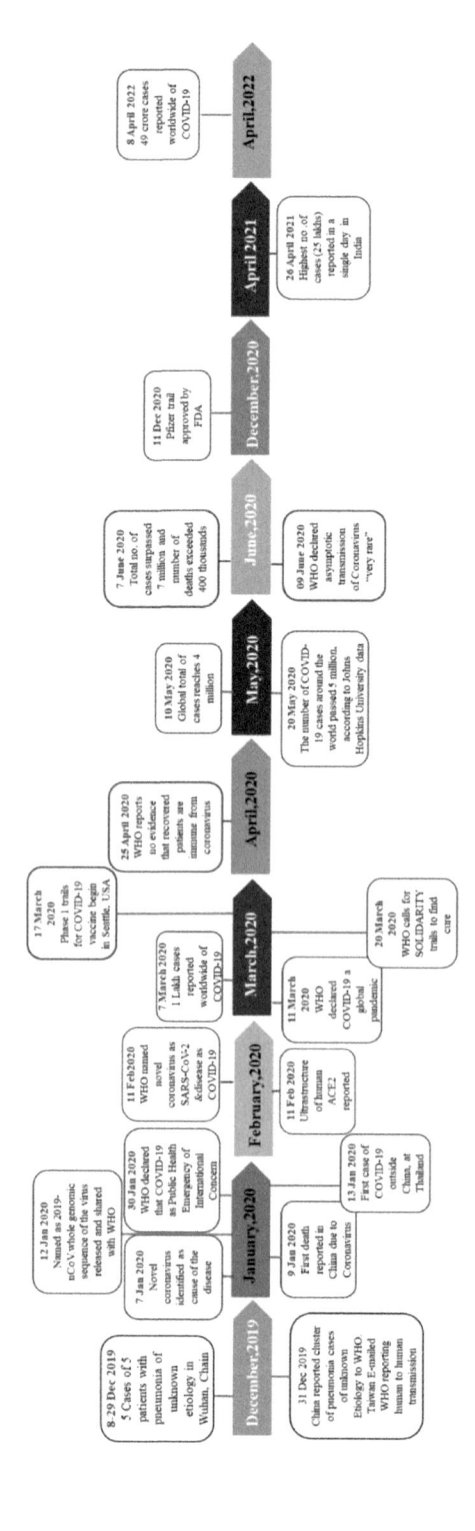

FIGURE 2.1   Timeline of various historic events reported for COVID-19.

that, under an electron microscope, resemble a crown. Several structural and nonstructural proteins are encoded by the coronavirus genome (**Figure 2.2a**). The structural proteins help with host infections, viral assembly, membrane fusion, viral particle release, and morphogenesis, among many other functions [14], while the nonstructural

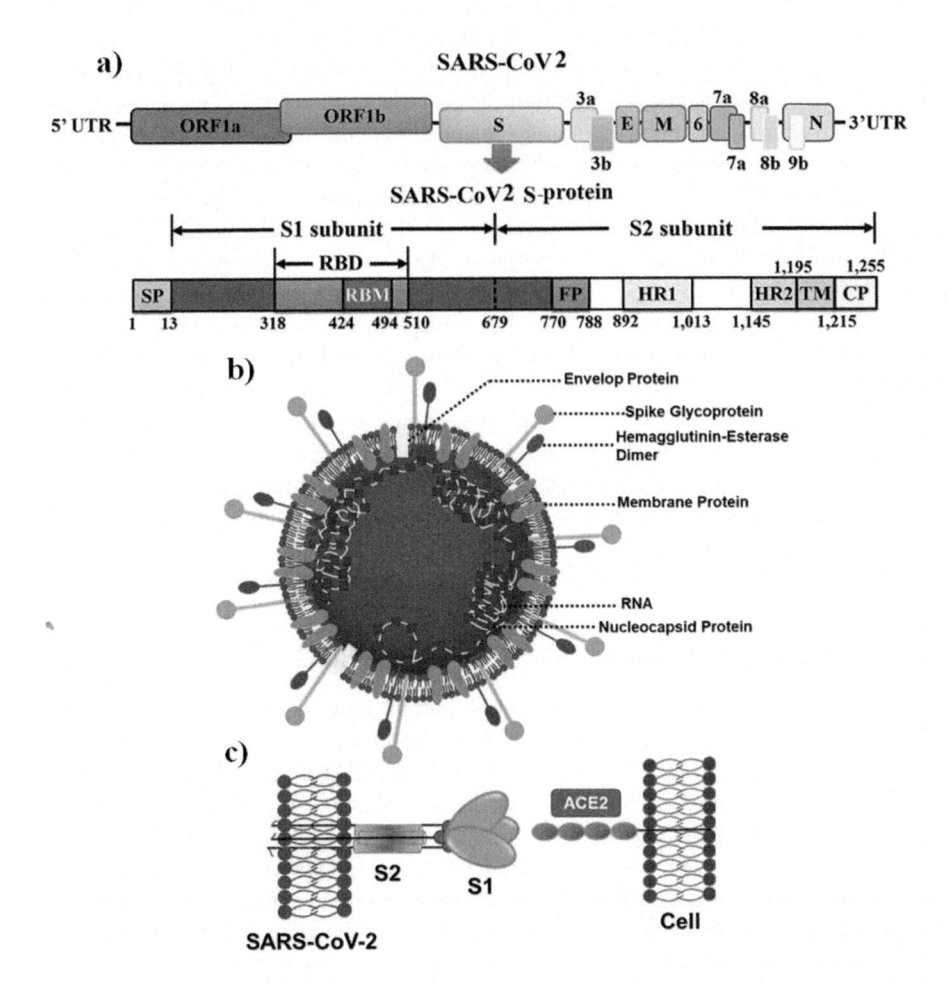

FIGURE 2.2 (a) COVID-19's genomic structure and several functional regions are depicted schematically. COVID-19's RNA contains two big genes, ORF1a and ORF1b, which code for 16 nonstructural proteins (nsp1–nsp16). E, M, S, and N structural proteins are encoded by the structural genes. The accessory genes are represented by different tones of green color. The structure of S protein is shown under the genomic organization. S1 and S2 subunits constitute the S protein. With a dotted line, the cleavage position of S1 and S2 is emphasized. The S protein contains the signal peptide (SP), the receptor-binding domain (RBD), the fusion protein (FP), the heptad repeat (HR), the transmembrane domain (TM), and the cytoplasmic domain (CP). (b) COVID-19 structure exhibiting multiple structural proteins as well as a single conventional positive-sense viral RNA coupled with nucleocapsid proteins. (c) Interactions of the spike proteins with the host cell receptor ACE2, which is important for viral binding to host cell and their fusion, which allows virus entrance into the host cell.

proteins help in its replication and transcription [15]. Nucleocapsid protein (N), membrane protein (M), envelop protein (E), and spike proteins (S) are the four structural proteins that make up the virion (S) (**Figure 2.2b**) [16]. Within the nuclear capsid, the **N protein** wraps the viral single-stranded RNA genome. This protein binds to viral RNA, allowing the virus to take over the host cell and turn it into a virus factory. **M protein** is the most abundant transmembrane protein found in the virion envelop that is assumed to be the coronavirus structure assembly's major controller. The **E protein**, with 76–109 amino acids, is a tiny part of the viral structure yet plays a critical role in virus structural assembly, membrane permeability, and virus–host cell interactions [13]. S proteins are type I transmembrane proteins with a clove shape and three sections: big ectodomain, a single-pass transmembrane, and an intracellular tail. The S protein's ectodomain is made up of two subunits: the S1 subunit, which has a receptor-binding domain (RBD), and the S2 membrane fusion subunit. The recognition of host cell surface receptors by the RBD is the starting point in viral infection, and coronavirus spikes' binding interactions with host receptors are essential for host range and cross-species transmission. MERS-CoV binds to dipeptidyl peptidase-4, HCoV-229E binds to human aminopeptidase N, HCoV-NL63 binds to angiotensin-converting enzyme 2 (ACE2), and SARS-CoV binds to angiotensin-converting enzyme 2 (ACE2) (**Figure 2.2c**). The hemagglutinin-esterase dimer (HE) has been discovered on the virion's surface. The HE protein plays a vital role in virus entrance to the host cell and is thus important for natural host–cell infection [17]. Using state-of-the-art cryo-EM techniques, Wall et al. and Wrapp et al. have revealed the complete structure of SARS-CoV-2 spike proteins in both open as well as closed conformations [18, 19].

2.2   MODE OF ACTION OF SARS-COV-2 VIRUS

The coronavirus's strategy of action in the human body consists of several phases [13]. The coronavirus spike proteins initially detect and adhere to the angiotensin-converting enzyme 2 (ACE2) receptors on the host cell structure (Figure 2.3). In two regions positioned at the interface of the S1 and S2 subunits, the viral S protein is accessible to proteolytic digestion by proteases enzymes, namely, trypsin and furin. The fusion peptide is released once the S2 domain is cleaved, which activates the membrane fusion process. Endocytosis is the process of a virus being ingested inside a human cell. The endosome then opens and releases the virus into the cytoplasm in the second stage. Proteasomes are involved in the uncoating of the viral nucleocapsid which results in the release of virus RNA into the cytoplasm of the host cell. New copies of viral genetic material are formed in the third phase, owing to replication and transcription processes facilitated by the replication–transcription complex (RTC), which engages host ribosomes. The structural proteins of the viruses M, S, and E are generated in the cytoplasm and then implanted into the endoplasmic reticulum before being moved to the endoplasmic reticulum–Golgi intermediate compartment in the fourth stage (ERGIC). Nucleocapsids are formed in the cytoplasm by the encapsulation of the replicated genome by N protein, which then assemble into new virions within the ERGIC membrane. In the last phase, the newly created virion is exported to the cell membrane of the infected cell via smooth-walled vesicles and subsequently secreted outside the cell via exocytosis to infect healthy cells. Meanwhile, the stress of producing new virions causes cell death in the endoplasmic reticulum.

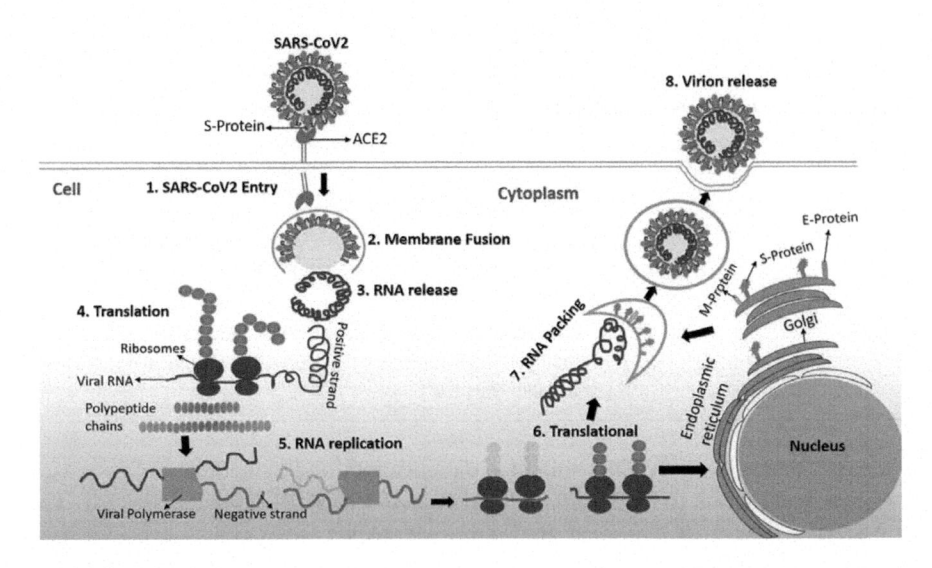

FIGURE 2.3 The schematic showing the mechanism of COVID-19 infectivity and multiplication in the human cells.

2.3 MODE OF TRANSMISSION OF SARS-COV-2 VIRUS

The SARS-CoV-2 virus is spread via fomites and droplets from an infected person to an uninfected one. This virus can be spread by indirect contact, such as by coming in contact to the contaminated surfaces and personal protective equipment (PPEs). SARS-aerosol CoV-2's nature was explored recently by quantifying viral RNA load in aerosol in two Wuhan hospitals, which revealed that this virus has a strong potential for aerosol transmission. As a result of polluted aerosol created during interventional procedures, there is a greater risk of this illness spreading in healthcare institutions. The fact that COVID-19 spreads mostly by airborne transmission is widely known now [20]. Other investigations have found fecal–oral transmission (virus found in a rectal swab) [21], disseminated across the intestinal tract [22], communicable through contaminated urine [23], is also responsible for COVID-19 spread. In context of the transmission of virus to the fetus from the infected mother, one research found that if the mother got infected in the later stage of pregnancy, there is no transfer of virus to the newborn [24]. However, additional investigations have proven vertical transmission of the SARS-CoV-2 virus, with newly born children displaying a range of clinical issues as a result of the viral infection, such as neurological impairment, elevated cytokine levels, and so on [25].

2.4 CLINICAL CHARACTERISTICS OF COVID-19

Fever and a dry cough were the most prevalent COVID-19 symptoms. In addition, the majority of the patients had bilateral pneumonia. People in old age and who have a medical history are more likely to become infected with this disease [24]. Leucopenia and lymphopenia were seen in the infected individuals. In comparison

TABLE 2.1

Clinical Symptoms of COVID-19 in Three Different Severity Levels

Severity Level of COVID-19	Clinical Manifestations
Mild	Fever, dry cough, fatigue, no pneumonia, or mild pneumonia
Severe	Blood oxygen saturation level ≤93%, respiratory frequency ≥30 min^{-1}; within 24 to 48 h, arterial oxygen partial pressure to fraction of inspired oxygen ratios <300, and/or lung infiltrate >50%; ICU needed
Critical	Acute respiratory distress syndrome, respiratory failure, sepsis, multiple organ damage, coagulation dysfunction, and excessive metabolic acidosis

to non-ICU patients, COVID-19-infected patients admitted to the ICU had increased plasma levels of IL2, IL7, IL10, GSCF, IP10, MCP1, MIP1A, and TNF [26]. The COVID-19 disease is classified into three degrees based on its severity: critical, severe, and moderate. The clinical manifestations of COVID-19 are shown in Table 2.1 in three severity categories.

REFERENCES

1. Wang M-Y, Zhao R, Gao L-J, Gao X-F, Wang D-P, Cao J-M. SARS-CoV-2: structure, biology, and structure-based therapeutics development. Frontiers in Cellular and Infection Microbiology. 2020;10.
2. Sun P, Lu X, Xu C, Sun W, Pan B. Understanding of COVID-19 based on current evidence. 2020;92:548–51.
3. Hu B, Guo H, Zhou P, Shi Z-L. Characteristics of SARS-CoV-2 and COVID-19. Nature Reviews Microbiology. 2021;19:141–54.
4. WHO. WHO coronavirus (COVID-19) dashboard. WHO; 2022.
5. Gorbalenya AE, Baker SC, Baric RS, de Groot RJ, Drosten C, Gulyaeva AA, Haagmans BL, Lauber C, Leontovich AM, Neuman BW, Penzar D, Perlman S, Poon LLM, Samborskiy DV, Sidorov IA, Sola I, Ziebuhr J. The species severe acute respiratory syndrome-related coronavirus: classifying 2019-nCoV and naming it SARS-CoV-2. Nature Microbiology. 2020;5:536–44.
6. Tyrrell DA, Bynoe ML. Cultivation of viruses from a high proportion of patients with colds. Lancet (London, England). 1966;1:76–7.
7. Lau SK, Woo PC, Yip CC, Tse H, Tsoi HW, Cheng VC, Lee P, Tang BS, Cheung CH, Lee RA, So LY, Lau YL, Chan KH, Yuen KY. Coronavirus HKU1 and other coronavirus infections in Hong Kong. Journal of Clinical Microbiology. 2006;44:2063–71.
8. van der Hoek L, Pyrc K, Jebbink MF, Vermeulen-Oost W, Berkhout RJ, Wolthers KC, Wertheim-van Dillen PM, Kaandorp J, Spaargaren J, Berkhout B. Identification of a new human coronavirus. Nature Medicine. 2004;10:368–73.
9. Holmes KV. SARS-associated coronavirus. The New England Journal of Medicine. 2003;348:1948–51.
10. Zaki AM, van Boheemen S, Bestebroer TM, Osterhaus AD, Fouchier RA. Isolation of a novel coronavirus from a man with pneumonia in Saudi Arabia. The New England Journal of Medicine. 2012;367:1814–20.
11. Gralinski LE, Menachery VD. Return of the coronavirus: 2019-nCoV. Viruses. 2020;12:135.

12. Li F. Structure, function, and evolution of coronavirus spike proteins. Annual Review of Virology. 2016;3:237–61.
13. Boopathi S, Poma AB, Kolandaivel P. Novel 2019 coronavirus structure, mechanism of action, antiviral drug promises and rule out against its treatment. Journal of Biomolecular Structure and Dynamics. 2021;39:3409–18.
14. Siu YL, Teoh KT, Lo J, Chan CM, Kien F, Escriou N, Tsao SW, Nicholls JM, Altmeyer R, Peiris JSM, Bruzzone R, Nal B. The M, E, and N structural proteins of the severe acute respiratory syndrome coronavirus are required for efficient assembly, trafficking, and release of virus-like particles. Journal of Virology. 2008;82:11318–30.
15. Snijder EJ, Decroly E, Ziebuhr J. Chapter three—the nonstructural proteins directing coronavirus RNA synthesis and processing. Advances in Virus Research. 2016;96:59–126.
16. Mittal A, Manjunath K, Ranjan RK, Kaushik S, Kumar S, Verma V. COVID-19 pandemic: insights into structure, function, and hACE2 receptor recognition by SARS-CoV-2. PLOS Pathogens. 2020;16:e1008762.
17. Lissenberg A, Vrolijk MM, Vliet ALWV, Langereis MA, Groot-Mijnes JDFD, Rottier PJM, Groot RJD. Luxury at a cost? Recombinant mouse hepatitis viruses expressing the accessory hemagglutinin esterase protein display reduced fitness in vitro. Journal of Virology. 2005;79:15054–63.
18. Walls AC, Park Y-J, Tortorici MA, Wall A, McGuire AT, Veesler D. Structure, function, and antigenicity of the SARS-CoV-2 spike glycoprotein. Cell. 2020;181:281–92.e6.
19. Wrapp D, Wang N, Corbett KS, Goldsmith JA, Hsieh C-L, Abiona O, Graham BS, McLellan JS. Cryo-EM structure of the 2019-nCoV spike in the prefusion conformation. Science. 2020;367:1260–3.
20. Jayaweera M, Perera H, Gunawardana B, Manatunge J. Transmission of COVID-19 virus by droplets and aerosols: a critical review on the unresolved dichotomy. Environmental Research. 2020;188:109819.
21. Xu Y, Li X, Zhu B, Liang H, Fang C, Gong Y, Guo Q, Sun X, Zhao D, Shen J, Zhang H. Characteristics of pediatric SARS-CoV-2 infection and potential evidence for persistent fecal viral shedding. Nature Medicine. 2020;26:502–5.
22. Zhou J, Li C, Liu X, Chiu MC. Infection of bat and human intestinal organoids by SARS-CoV-2. Nature Medicine. 2020;26:1077–83.
23. Sun J, Zhu A, Li H, Zheng K, Zhuang Z, Chen Z, Shi Y, Zhang Z, Chen SB, Liu X, Dai J, Li X, Huang S, Huang X, Luo L, Wen L, Zhuo J, Li Y, Wang Y, Zhang L, Zhang Y, Li F, Feng L, Chen X, Zhong N, Yang Z, Huang J, Zhao J, Li YM. Isolation of infectious SARS-CoV-2 from urine of a COVID-19 patient. Emerging Microbes & Infections. 2020;9:991–3.
24. Chen N, Zhou M, Dong X, Qu J, Gong F, Han Y, Qiu Y, Wang J, Liu Y, Wei Y, Xia J, Yu T, Zhang X, Zhang L. Epidemiological and clinical characteristics of 99 cases of 2019 novel coronavirus pneumonia in Wuhan, China: a descriptive study. Lancet (London, England). 2020;395:507–13.
25. Chen H, Guo J, Wang C, Luo F, Yu X, Zhang W, Li J, Zhao D, Xu D, Gong Q, Liao J, Yang H, Hou W, Zhang Y. Clinical characteristics and intrauterine vertical transmission potential of COVID-19 infection in nine pregnant women: a retrospective review of medical records. Lancet (London, England). 2020;395:809–15.
26. Huang C, Wang Y, Li X, Ren L, Zhao J, Hu Y, Zhang L, Fan G, Xu J, Gu X, Cheng Z, Yu T, Xia J, Wei Y, Wu W, Xie X, Yin W, Li H, Liu M, Xiao Y, Gao H, Guo L, Xie J, Wang G, Jiang R, Gao Z, Jin Q, Wang J, Cao B. Clinical features of patients infected with 2019 novel coronavirus in Wuhan, China. Lancet (London, England). 2020;395:497–506.

3 2D Materials
Synthesis and Properties

Pradip Kumar

The interest in two-dimensional (2D) materials is continuously increasing after graphene isolation from graphite using mechanical exfoliation in 2004 [1]. Graphene is a single layer of sp^2 hybridized carbon atoms arranged in a honeycomb lattice structure. Single-layer graphene exhibits ultimate carrier mobility (μ ~10,000 cm^2 V^{-1} s^{-1} at RT), excellent thermal conductivity (~3,000–5,000 W/mK), high Young's modulus (~1 TPa), ultrahigh surface area (~2,630 m^{2}g^{-1}), and excellent optical (absorb only 2.3% of white light) properties. The unique properties of graphene inspired the research community to develop more and more new 2D materials. After that, many 2D materials, including monoelemental 2D materials (borophene, phosphorene, silicene, etc.); transition metal dichalcogenide (TMDs), such as WS_2, MoS_2, $MoSe_2$, and WSe_2; transition metal carbides/nitrides (MXenes); hexagonal boron nitride (h-BN); covalent oxide frameworks (COFs); metal oxide frameworks (MOFs); layered double hydroxides (LDHs); black phosphorous (BP); and oxides, have been explored (Figure 3.1a) [2]. Among 2D materials, graphene, MXenes, and MoS_2 are the most studied materials for various applications, including electronics, sensors, energy storage, energy conversion, thermal management, and electromagnetic interface shielding.

The development of ultrathin 2D materials has been demonstrated by various experimental processes, including micromechanical cleavage, liquid-phase exfoliation, ion-intercalated exfoliation, wet-chemical synthesis, and chemical vapor deposition methods (Figure 3.1b) [3]. The properties of the developed materials are highly dependent on the synthesis methods. For example, CVD-produced graphene is highly electrical and thermally conducting. In contrast, liquid-phase exfoliated (Hummer's method) graphene oxide (GO) is insulating in nature due to the presence of oxygen-containing functional groups on its basal plane and surface. GO can be conductive upon reduction of functional groups by chemical or thermal reduction method. However, the properties of reduced graphene oxide are not as good as those of CVD-produced graphene. Similarly, the synthesis method is very important for the development of other 2D materials. The synthesis process can be optimized by various techniques, including microscopic techniques (e.g., scanning electron microscopy, transmission electron microscopy, atomic-force microscopy, scanning tunnelling microscopy), Raman, FTIR, and X-ray diffraction [4, 5]. In this chapter, recent advancements in 2D materials synthesis and their properties are discussed.

 DOI: 10.1201/9781003316381-3

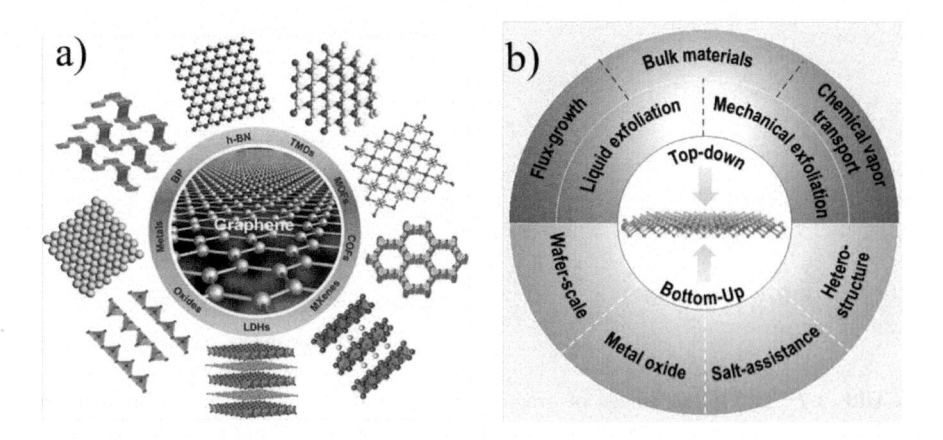

FIGURE 3.1   (a) Illustration of the family of 2D materials like graphene, TMDs, h-BN, BPs, metals, oxides, LDHs, MXenes, COFs, and MOFs. (b) Synthesis overview of 2D materials using various top-down and bottom methods.

Source: (a) Reproduced with permission from ref. [6]. (b) Reproduced with permission from ref. [3].

3.1   SYNTHESIS AND PROPERTIES

3.1.1   GRAPHENE

Graphene has been synthesized *via* both top-down and bottom-up approaches. Nowadays, graphene production is well established and is commercially available. Both top-down approaches, such as micromechanical cleavage (MC) and liquid-phase exfoliation, and bottom-up approaches, like CVD and molecular beam epitaxy, are well established. Large-scale production of graphene can be obtained using liquid-phase exfoliation and the CVD process. In this section, graphene synthesis method and its properties are discussed.

3.1.1.1   Dry Exfoliation

Dry exfoliation is the process where layered materials split into an atomically thin layer by applying mechanical, electromagnetic, or electrostatic force under an appropriate environment. Graphene was first synthesized by the micromechanical cleavage (MC) method using adhesive tape. MC method can produce single- and multilayer graphene (Figure 3.2a). The number of layers can be easily identified by Raman spectroscopy (Figure 3.2b). This method is good for the synthesis of high-quality graphene for fundamental studies and not for bulk production. The decoupled single-layer graphene (SLG) on the surface of bulk graphite can achieve carrier motilities up to 10^7 cm^{2}V^{-1}s^{-1} at RT [7], while suspended SLGs nobilities can reach up to 10^6 cm^{2}V^{-1}s^{-1} [8]. The thermal conductivity of suspended SLGs was reported up to $5.30 \pm 0.48 \times 10^3$ W/mK at room temperature [9]. The superb thermal conductivity of SLGs is mainly due to the acoustic phonons with a negligible electronic contribution.

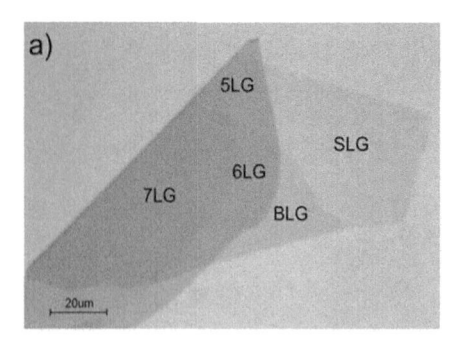 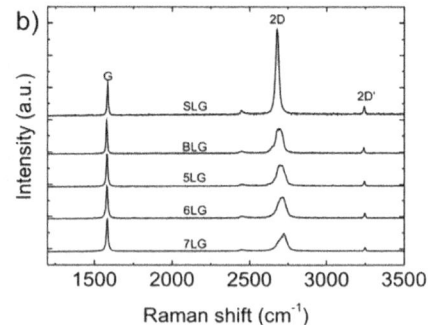

FIGURE 3.2   (a) Optical image of graphene flakes of different thicknesses produced by mechanically exfoliation method and (b) corresponding Raman spectra for different layers.

Source: Reproduced with permission from ref. [12].

Single-layer graphene showed Young's modulus up to 1 TPa and intrinsic tensile strength of 130 GPa [10]. Interestingly, monolayer graphene is found to absorb only 2.3% of incident white light [11]. These unique properties of graphene make it one of the best materials for various potential applications.

Graphene was further synthesized using an anodic bonding technique. In this method, first, graphite is pressed onto the glass substrate, and then a voltage of 0.5–2 kV is applied between the metal contact and graphite, followed by 10–20 minutes of heating at ~200°C [13, 14]. This technique can produce graphene flakes up to ~1 mm in width. The size and number of graphene layers can be controlled by the applied voltage and temperature [14]. Dry exfoliation of graphite can also be achieved using the laser ablation technique. In this technique, a laser pulse is used to exfoliate the graphite flakes. Single- or multilayer graphene can be obtained by changing the laser energy density [15]. The result shows that the number of graphene layers decreases by increasing the laser energy density.

3.1.1.2   Liquid-Phase Exfoliation of Graphite

Graphene/graphene oxide is extensively synthesized using liquid-phase exfoliation of graphite under appropriate conditions. In this process, first, graphite is dispersed in a solvent and then exfoliated and then washed/purified. The purification of exfoliated graphite generally requires ultracentrifugation. Single- and multilayer graphene sheets can be produced by exfoliation of graphite into water and organic solvents by ultrasonication [16, 17]. The graphite dispersion in a solvent depends on the interfacial tension between graphite and solvent. The solvents with a low surface tension of γ ~40 mN/m are the best for graphite and graphene dispersion [16]. Despite the good dispersibility of graphite in solvents (i.e., DMF and NMP), γ ~40 mN/m has some disadvantages, like their toxic nature [18]. Also, solvent removal after exfoliation is difficult due to their high boiling points (>450 K). Graphite dispersion and exfoliation can be obtained in low-boiling-point solvents, like acetone and isopropanol [19]. Graphite dispersion and exfoliation in water are difficult due to its higher surface tension of γ ~72 mN/m [20]. However, the dispersion and stability of graphene flakes in water can be achieved by surfactants or polymers [21]. This liquid-phase exfoliation of graphite can produce graphene flakes size ranging from a nanometer to a few micrometers. The production yield of single-layer graphene flakes depends on the size of graphite flakes and solvents.

Another method of graphene synthesis is the exfoliation of graphite oxide or graphite-intercalated compounds. In this process, first, graphite is oxidized in the presence of a suitable oxidizing agent, followed by exfoliation and purification. In 1859, for the first time, Brodie noticed graphite oxidation by investigating its reactivity in potassium chlorate ($KClO_3$) and fuming nitric acid (HNO_3) [22]. The overall weight of reacted graphite was increased and was composed of carbon (61.04), hydrogen (1.85), and oxygen (37.11). The molecular formula of oxidized graphite was $C_{2.19}H_{0.80}O_{1.00}$. Later on, Brodie's $KClO_3$-fuming HNO_3 graphite oxidation process was improved by obtaining C: O ~2:1 [23]. Nearly 100 years after Brodie, Hummers and Offeman modified the process of graphite oxidation to avoid explosive by-products, as in the case of $KClO_3$ [24]. They developed the process of graphite oxidation in a mixture of sulfuric acid, potassium permanganate, and sodium nitrate for graphite oxidation. In this method, the sp^2-bonded network of graphite was disrupted and introduced oxygen-containing functional groups on its basal plane and edges, as shown in Figure 3.3a [12]. Later on, Hummer's method was modified by varying the ratio of H_2SO_4, $KMnO_4$, and HNO_3. This method can produce graphene oxide

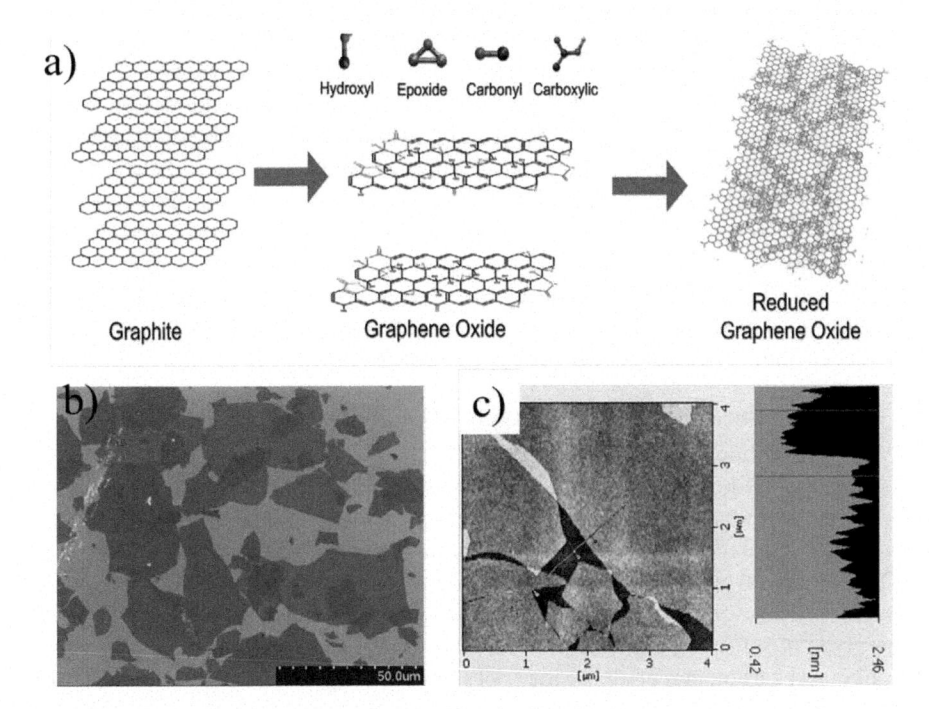

FIGURE 3.3   (a) GO synthesis by exfoliation of graphite in the presence of strong acid and reduction of GO sheets. (b) FESEM and (c) AFM images of monolayer GO sheets.

Source: (a) Reproduced with permission from ref. [12]. (b, c) Reproduced with permission from ref. [25, 26].

(GO) of various sizes ranging from a few micrometers to 100 μm, depending on the size of graphite flakes. The typical thickness of single-layer GO is ~1 nm, which is three times higher than bare graphene. This higher thickness of single-layer GO sheets is due to the presence of oxygen-containing functional groups on their surface. Figure 3.3b,c shows the representative FESEM and AFM images of monolayer GO sheets [25, 26]. The mechanism of GO synthesis has been well described in the literature [27].

Graphene oxide obtained by graphite oxidation and exfoliation method is highly dispersible and stable in polar as well as nonpolar solvents due to the presence of functional groups (hydroxyl, carbonyl, epoxide, and carboxylic). The aqueous dispersion of 2D graphene oxide exhibits a stable liquid crystalline (LC) phase. Pasquali et al. reported the first experimental evidence for the LC phase of colloidal graphene [28]. In their work, spontaneous exfoliation of graphite into graphene, and their colloidal dispersion, was obtained in a harsh superacid, such as chlorosulfonic acid. A typically birefringent nematic schlieren texture of dispersed graphene was obtained at a concentration of ~2 mg/mL. Highly dispersible graphene is in demand for various applications, including composites and flexible electronics, but acid-assisted dispersion is not recommended for practical uses. Kim et al. reported the first study of graphene oxide liquid crystal in aqueous dispersion. Aqueous GO dispersion exhibits an inhomogeneous dark chocolate-milk-like wavy appearance, which indicates the formation of the nematic LC phase [29]. The LC phase of GO dispersion is further confirmed by polarized optical microscopy and scanning electron microscopy. The properties of GO liquid crystals and their potential applications are described in the literature [30, 31].

Pristine GO has an insulating nature due to the presence of functional groups on its surface. But GO can achieve pretty good electrical and thermal conductivity upon the reduction of oxygen-containing functional groups [32, 33]. The chemically or thermally reduced graphene oxide (rGO) exhibits good electrical and thermal conductivity, which is useful for the fabrication of thin films, foam, composites for electromagnetic interference shielding, thermal management, catalyst, and energy storage applications [34–38]. Further, the presence of functional groups makes the GO sheets more suitable for structural modification [39].

3.1.1.3 Chemical Vapor Deposition (CVD)

CVD is a traditional technique that has been used for the production of highly purified materials, like Si, Zr, Ti, and Ta. Currently, the CVD technique has been used for the mass production of various 2D materials, including graphene, transition metal dichalcogenides (TMDs), and boron nitride. In a typical CVD method, the growth of 2D materials involves the decomposition of hydrocarbons and nucleation of carbon atoms on the given substrate in a preselected condition, such as temperature, pressure, type of substrate, and catalysts. Methane (CH_4) and acetylene (C_2H_2) are commonly used carbon precursors in the CVD process. The various substrate, including Ni, Cu, Ir, SiO_2, and Fe, were used for graphene growth, but Cu is the most common substrate for graphene growth in CVD techniques. The CVD method is one of the widely used techniques for graphene growth. This method can produce very high-quality single-layer and multilayer graphene on the desired metal substrate. The quality of CVD-grown graphene is as good as the Scotch tape method–produced graphene. There are different types of CVD processes, including thermal, plasma-enhanced, hot wall, reactive, etc., that have been explored for graphene synthesis. The main difference in different CVD equipment is the type of precursor source.

For the first time, Johann et al. reported graphene growth on Ir(111) using the thermal CVD process [40]. In their attempt, low-pressure CVD allows high-quality single-layer graphene growth on Ir(111) substrate using an ethylene precursor. Graphene can be grown on Ir substrate by CVD due to its low carbon solubility. But graphene transfer on other substrates is difficult due to its chemical inertness. Further, graphene was grown on Ni, Co, and Cu substrates by CVD [41–43]. Graphene grown on Ni and Co substrate produced few-layer graphene (FLG), not single-layer graphene (SLG). Li *et al.* reported the first single-layer large-area graphene (~centimeter size) on Cu substrate by the CVD process using a methane precursor [44]. As deposited, SLG film has shown electron mobilities as high as 4,300 $cm^2V^{-1}S^{-1}$. In 2010, roll-to-roll production of graphene on Cu foil was reported up to 30 inch [45]. The as-deposited graphene film on Cu shows low sheet resistance ~125 $\Omega\square^{-1}$ with 97.4% optical transparency. The transfer of CVD-grown graphene to other substrates is the most challenging task. Graphene films (~50 cm) with μ ~7,350 $cm^2V^{-1}S^{-1}$ at 6 K were transferred *via* roll-to-roll (R2R) process, as shown in Figure 3.5. Graphene was transferred on a transparent conductive film (GTCF) with a low sheet resistance ~150 $\Omega\square^{-1}$ using photocurable epoxy resin. Graphene on Cu was transferred to SiO_2 substrate with μ ranging from ~16,400 to ~25,000 $cm^2V^{-1}S^{-1}$ and h-BN substrate with μ ranging from ~27,000 to ~45,000 $cm^2V^{-1}S^{-1}$ at room temperature [46]. After that, many other metal substrates, including ruthenium and platinum, were employed for graphene growth up to lateral size (lateral) of centimeters [47]. A large-size graphene/PET sandwich structure of 40 inches in diagonal was obtained via R2R process using atmospheric CVD (Figure 3.4) [48].

The plasma-enhanced CVD (PECVD) method is the most common and inexpensive technique used for graphene synthesis. In this technique, plasma reaction with gaseous

FIGURE 3.4 Graphene growth design via R2R process at atmospheric CVD.

Source: Reproduced with permission from ref. [48].

precursors lowers the deposition temperature in comparison to thermal CVD. However, plasma can induce some defects during material growth. Terasawa et al. reported graphene growth on Cu by plasma-enhanced CVD at low temperature (500°C) [49]. Graphene growth was optimized by changing plasma power, substrate temperature, and gas pressure. Single-layer graphene was obtained at high substrate temperature, but the film quality was not good, as in the case of exfoliated or thermal CVD graphene. Later on, the growth of vertical SLGs and FLGs on various substrates, including SiO_2 and carbide-forming substrates, was reported by microwave PECVD [4].

CVD graphene has been successfully grown on many substrates; however, its transfer on the target substrate is a challenging task, which is needed for electronics and sensor applications. Along with the advancement in CVD-grown graphene, progress in transfer methods on different substrates is also developed. Graphene transfer by continuous transfer *via* R2R process and polymer-assisted method has been well established. Polymer-assisted method is the most popular for transferring graphene grown on a transition metal substrate to the target substrate. In this process, a poly(methyl-methacrylate) (PMMA) layer is coated onto the graphene, and then the metal substrate is completely etched away using a suitable etchant. Subsequently, PMMA-coated graphene is transferred onto the target substrate, followed by the removal of the PMMA layer using suitable solvents. This process is simple, but there is some PMMA residue left on the graphene surface, which slows down the charge carriers in graphene. Also, the solvent used for PMMA removal may cause some structural defects in the graphene surface. This polymer-assisted transfer process of graphene on the target substrate is simple and most popular. Further, graphene transfer on the required substrate is demonstrated in Figure 3.5 [50, 51].

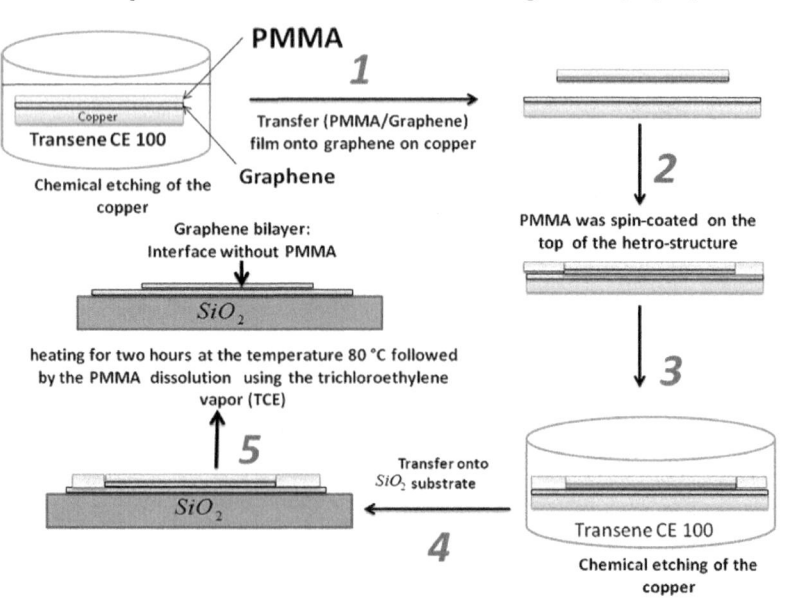

FIGURE 3.5   Illustration of PMMA-assisted CVD-grown graphene transfer process on a target substrate.

Source: Reproduced with permission from ref. [51].

3.1.2 OTHER TECHNIQUES

Graphene can be synthesized by substrate-based techniques, wherein a single layer of carbon atoms grows on single-crystal carbide (SiC) under vacuum graphitization. Graphene films were obtained by thermal decomposition of SiC annealed at high temperature (>1,000°C) under ultrahigh vacuum (UHV) [52]. The thermal decomposition of SiC results in the sublimation of the Si atoms, while the C-enriched surface rearranges to form graphene films. Graphene growth can be controlled by precise control of the Si sublimation [53]. Graphene grown on Si-face has lower mobility (~500 to ~2,000 $cm^2V^{-1}S^{-1}$) than C-face (~10,000 to ~30,000 $cm^2V^{-1}S^{-1}$). Tromp *et al.* reported the thermodynamics and kinetics of graphene growth on SiC(001) surface [54]. Recent advances in graphene growth on SiC have been reviewed elsewhere [55]. This technique of graphene growth is best suited for light-emitting devices and high-frequency electronics applications. The SiC substrate is an established material for power electronics, so graphene growth directly on the SiC substrate is good for electronics applications. Graphene on SiC has been demonstrated for transistor applications [56]. However, large-scale production is the main concern of the epitaxial growth method.

Further, graphene synthesis was also reported by unzipping CNTs and MWCNTs. This method mainly produces graphene nanoribbons (GNRs) of typical width of ≤50 nm and an aspect ratio of ≥10. Amanda et al. reported GNRs synthesis by opening MWCNTs longitudinally in the presence of potassium permanganate and sulfuric acid [57]. The yield of GNRs synthesis using the unzipping of CNTs approach is nearly 100%, but this process introduces some functional groups of GNRs surface, which can be reduced by the thermal or chemical reduction process. Later on, the crystalline graphene structure was synthesized by electrochemical unzipping of heteroatom-doped (nitrogen) CNTs [58]. The study revealed that CNTs wall unzipping was initiated by substitutional pyridinic nitrogen (Np) dopant. This dopant-specific unzipping of CNTs at moderate potential produces crystalline graphene nanostructures with fine control and well-defined edge configuration.

3.2 TRANSITION METAL CARBIDES/NITRIDES (MXENES)

Two-dimensional (2D) transition metal carbides and nitrides or carbonitrides referred to as MXenes were first reported in 2011 [59]. Since then, advancement in the synthesis of various MXenes for their wide range of applications has been demonstrated. Currently, MXenes are one of the most versatile 2D materials for scientists and related industries. The general chemical formula of MXenes is $M_{n+1}X_nT_x$, where M is an early-transition metal (Ti, Sc, Y, Zr, Hf, V, Nb, Ta, Cr, Mo, W), X is carbon or nitrogen, T_x stands for surface functional groups (O, OH, F, and Cl), and $n = 1–4$ [60]. There are more than 30 stoichiometric compositions, and 100 stoichiometric MXenes structures have been proposed theoretically. Studies found that carbon-containing MXenes are more stable than nitrogen-containing MXenes due to the lower cohesive energy of nitrides. Depending on the chemical composition, MXenes have tunable properties, like electrical, thermal, mechanical, and surface area, which make them unique materials for versatile applications [5, 61, 62].

In a typical synthesis, MXenes are synthesized topochemical from their parent MAX phase *via* selective etching of the A layer. The first MXene, $Ti_3C_2T_x$, was synthesized by immersing the Ti_3AlC_2 MAX phase in the 50% HF solution for 2 h at RT [59]. After that, the reacted MAX-phase dispersion was washed with DI and then ultrasonicated and centrifuged. The resulted supernatant was collected as single-layer $Ti_3C_2T_x$ flakes. During the etching process, some functional groups, such as OH and F, were also attached to the transition metal. The suggested mechanism of $Ti_3C_2T_x$ MXenes is described as follows:

$$Ti_3AlC_2 + 3HF = AlF_3 + 1.5H_2 + Ti_3C_2 \tag{1}$$
$$Ti_3C_2 + 2H_2O = Ti_3C_2(OH)_2 + H_2 \tag{2}$$
$$Ti_3C_2 + 2HF = Ti_3C_2F_2 + H_2 \tag{3}$$

The synthesis of MXenes depends on various parameters, like the size of the MAX phase, etchant, and M–A bond strength. In the MAX phase, the M–A bond is weaker than the M–X bond, which favors the selective etching of the A element upon intercalation of the suitable etchant in the MAX phase. MXenes can be synthesized from MAX phases and non-MAX phases (Figure 3.6). As synthesized MXenes, flakes can be easily dispersed in water. Alhabeb et al. reported detailed guidelines for the synthesis and processing of $Ti_3C_2T_x$ MXenes using different etchants and delamination methods [63]. They have optimized synthesis conditions, including HF concentration, reaction time, different etchants, and delamination methods.

The most common etchant in MXenes synthesis is HF. But its toxic nature is a concern. Other etchants, including ammonium bifluoride (NH_4HF_2), a mixture of LiF and HCl, and fluoride salts (CaF_2, NaF, NH_4F, and FeF_3) mixtures with HCl, have been used for MXenes synthesis [64, 65]. The 2D $Ti_3C_2T_x$ MXene, synthesized using a mixture of 6 M HCl and LiF, was found to have fewer defects, with a larger size [66]. Further, $Ti_4N_3T_x$ MXenes were prepared using ternary eutectic molten fluoride salt (LiF, NaF, and KF) etchant [67]. As prepared, $Ti_4N_3T_x$ MXene flakes were found to be more inferior in crystallinity than $Ti_3C_2T_x$ MXenes, wherein fluoride salt was used as an etchant. MXene was also synthesized using fluorine-free alkaline etchant (NaOH) under given hydrothermal conditions [68]. In this approach, a concentrated NaOH (27.5 M) solution at a high temperature (270°C) was used for the conversion of $Al(OH)_3$ and $AlO(OH)$ into dissolvable $Al(OH)_4^-$. This method can produce multilayer MXene sheets with no HF. However, hydrothermal conditions, like high temperature and pressure, have a safety concern, which was resolved by using new tetramethylammonium hydroxide (TMAOH) etchant [69]. Similar to graphene, MXenes can be synthesized by the electrochemical method [70]. In electrochemical method, a certain voltage (e.g., 0.4–0.7 V for V_2AlC, 0.3 V for Ti_2AlC, and 0.6–1.0 V for Cr_2AlC) was applied to MAX phase to etch out the Al layers [71]. This is a simple process, but it is not suitable for bulk synthesis due to the long etching time. To date, different MXenes have been explored, but $Ti_3C_2T_x$ is the most studied MXenes used for various applications. Table 3.1 summarizes the synthesis of various MXenes using different methods [61].

Similar to graphene, some MXenes have outstanding electrical, thermal, and mechanical properties. The mechanical properties of MXenes were experimentally measured on single flakes. The Young's modulus of $Ti_3C_2T_x$ and $Nb_4C_3T_x$ MXenes

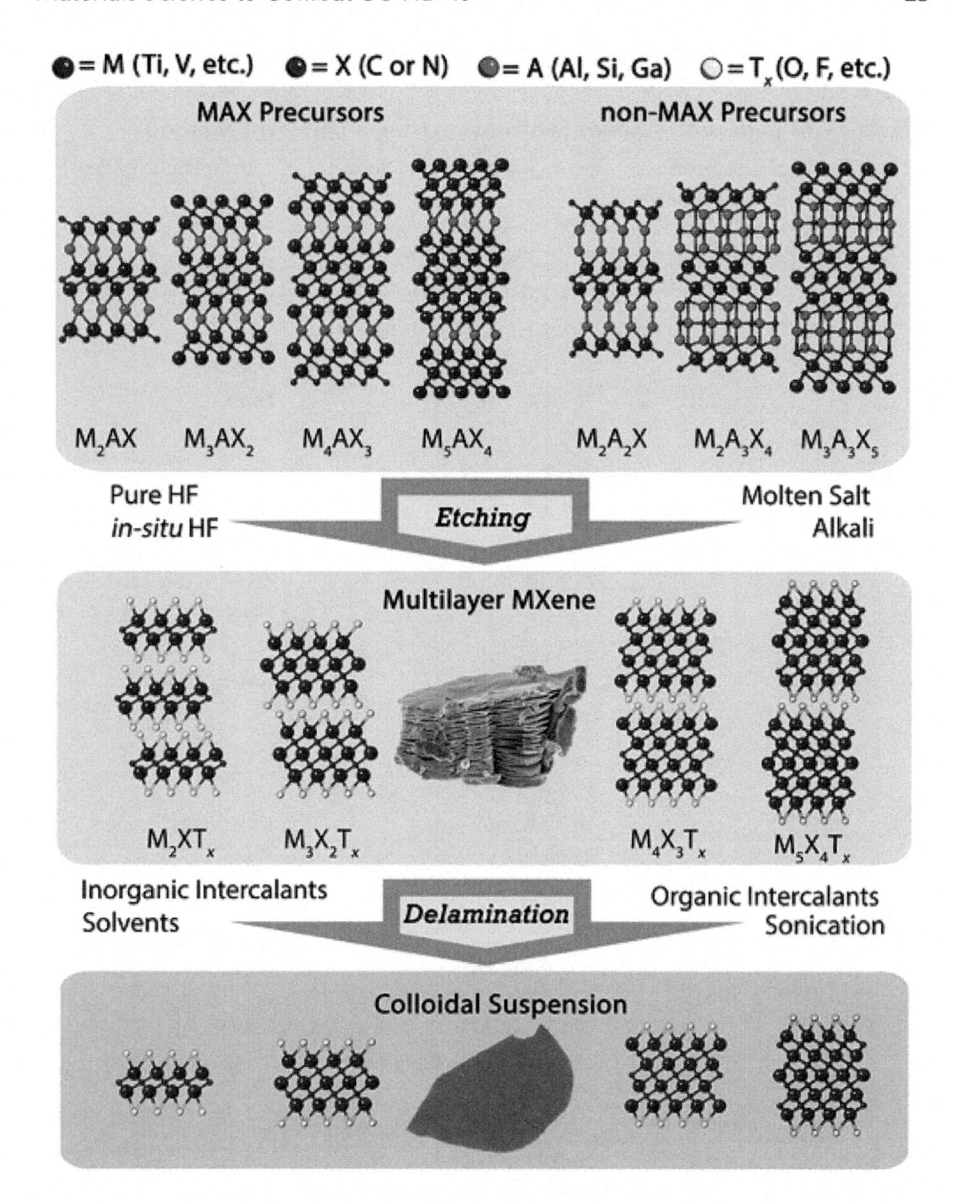

● = M (Ti, V, etc.)   ● = X (C or N)   ● = A (Al, Si, Ga)   ○ = T_x (O, F, etc.)

MAX Precursors   **non-MAX Precursors**

M_2AX M_3AX_2 M_4AX_3 M_5AX_4 M_2A_2X $M_2A_3X_4$ $M_3A_3X_5$

Pure HF   **Etching**   Molten Salt
in-situ HF   Alkali

Multilayer MXene

M_2XT_x $M_3X_2T_x$ $M_4X_3T_x$ $M_5X_4T_x$

Inorganic Intercalants   **Delamination**   Organic Intercalants
Solvents   Sonication

Colloidal Suspension

FIGURE 3.6 Selective etching of A layer from MAX and non-MAX phase precursor produces single-layer and multilayer MXenes and formed colloidal suspension.

Source: Reproduced with permission from ref. [5].

was experimentally measured to be 0.33 ± 0.03 TPa and 0.39 ± 0.02 TPa, respectively [72, 73]. The comparative electrical conductivity of freestanding or supported MXenes films is shown in Figure 3.7a. Among all MXenes, $Ti_3C_2T_x$ MXene exhibits the highest electrical conductivity of ~8,570 S cm^{-1}. Also, the intrinsic charge carrier density (~2×10^{21} cm^{-3}) and carrier mobility (~34 cm^{2}V^{-1} s^{-1}) of $Ti_3C_2T_x$ MXene are higher than those of Mo-based MXenes (~10^{20} cm^{-3} for $Mo_2Ti_2C_3T_x$ and ~10^{19} cm^{-3}

TABLE 3.1
Summary of Different MXenes Synthesis _Using a Different Method

S. No.	MXene	MAX Precursor	Synthesis Conditions	Delaminating Agent
1	$Ti_3C_2T_x$	Ti_3AlC_2	50% HF, 2 h	Sonication
2	$Ti_3C_2T_x$	Ti_3AlC_2	50% HF for 96 h	–
3	Ti_3C_2	Ti_3AlC_2	50% HF, 1 M NH_4HF_2	–
4	V_4C_3	V_4AlC_3	40% HF for 24 h and 96 h	–
5	$Ti_3C_2T_x$	Ti_3AlC_2	50% HF, 36 h	DMSO
6	$Ti_3C_2T_x$	Ti_3AlC_2	50% HF, 18 h	DMSO
7	Ti_2C	Ti_2AlC	10% HF, 18 h	–
8	Mo_2C	Mo_2GaC	~50% HF, 6.6 days	–
9	$Ti_3C_2T_x$	Ti_3AlC_2	50% HF, 18 h	Hydrazine monohydrate
10	Ti_3CNT_x	Ti_3AlCN	30% HF, 18 h	TBAOH
11	V_2CT_x	V_2AlC	48%HF, 92 h	TBAOH, choline hydroxide, n-butylamine
12	$Hf_3C_2T_x$	$Hf_3(Al(Si))_4C_2$	35% HF, 60 h	–
13	Nb_2C	Nb_2AlC	50% HF, 48 h	Isopropylamine
14	$Mo_2Ti_2C_3$	$Mo_2Ti_2AlC_3$	48–51% HF, 90 h	–
15	Mo_2TiC_2	Mo_2TiAlC_2	48–51% HF, 48 h	DMSO
16	$Ti_3C_2T_x$	Ti_3AlC_2	9 M HCl + 49% HF + DI water	LiCl
17	Ti_3C_2, Ti_2C	Ti_3AlC_2, Ti_2AlC	HCl + fluoride salt, 20 ml HCl	DMSO, NH_3, Urea, H_2O
18	Ti_3C_2	Ti_3AlC_2	9 g LiF + 9 M HCl	DMSO
19	V_2C	V_2AlC	2 g NaF + 40 ml HCl	–
20	Ti_2N	Ti_2AlN	6 g KF + 6 MHCl, 1 h	DMSO
21	Ti_2NT_x	Ti_2AlN	KF (0.59) + LiF (0.29) + NaF (0.12) and HCl	–
22	Ti_4N_3	Ti_4AlN_3	KF (0.59) + LiF (0.29) + NaF (0.12) and H_2SO_4	TBAOH
23	Ti_3C_2	Ti_3AlC_2	NH_4F (hydrothermal), 150°C, 24 h	–
24	Ti_3C_2	Ti_3AlC_2	NaOH (27.5 mol/L) at 270°C	–
25	Ti_2CT_x,	Ti_2AlC	Electrochemical method (0.3 V)	–
26	V_2CT_x and Cr_2CT_x	V_2AlC and Cr_2AlC	Electrochemical method (0.4–0.7 V for V_2AlC and 0.6–1.0 V for Cr_2AlC)	–

Note: –, value not available.
Source: Reproduced with permission from ref. [61].

for Mo_2TiC_2Tx) [74, 75]. Figure 3.7b shows the UV-visible spectra of Ti_2CTx and Ti_3C_2Tx MXenes samples. Both MXenes samples show electronic transitions and plasmonic effects. In the case of $Ti_3C_2T_x$, the plasmonic peak at 780 nm gives a green color from diluted aqueous Ti_3C_2Tx solution and a purple color from the film, while

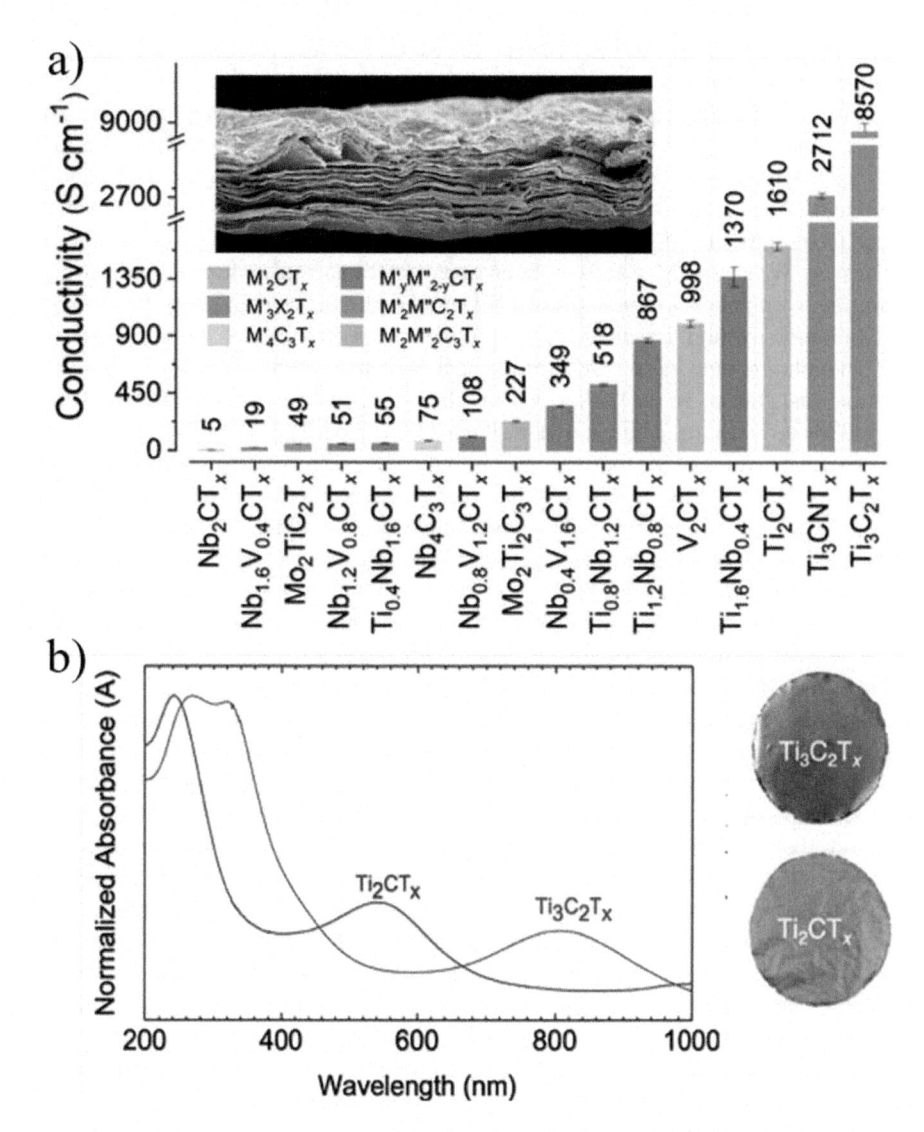

FIGURE 3.7 (a) Electrical conductivity of different MXene films. (b) Optical absorbance spectra of $Ti_3C_2T_x$ and Ti_2CT_x films along with their freestanding photos.

Source: (a) Reproduced with permission from ref. [76]. (b) Reproduced with permission from ref. [63].

a plasmonic peak around 550 nm was measured for thinner Ti_2CT_x MXene, which gives it a green color from the film and a vine color from the solution.

3.3 TRANSITION METAL DICHALCOGENIDES (TMDS)

TMDs have received significant research interest due to their excellent electrical, mechanical, and optical properties. Transition metal dichalcogenides (TMD_S) are a form of MX_2, where M is the transition metal element from groups 4 to 10 and X is the chalcogen element (S, Se, or Te). In general, TMD materials composed of groups 4–7 transition metal elements have a layered structure, while groups 8–10 transition metals form a non-layered structure. TMDs materials consist of a hexagonally packed layer of metal atoms *via* weak van der Waals forces. The metal atoms provide four electrons to fill the bonding states of TMDs such that the oxidation states of M and X are +4 and –2, respectively. The different combinations of transition metals and chalcogens form more than 40 TMD materials [77]. The most commonly studied TMDs, MoS_2, WS_2, WSe_2, and $MoSe_2$, have been explored for various potential applications.

Similar to graphene, a monolayer of TMD materials can be obtained by the mechanical exfoliation method. Desai et al. reported the Au-mediated exfoliation of large-size TMD monolayers onto various substrates, such as quartz and SiO_2/Si [78]. First, adhesive tape was attached to the bulk MX_2, and then Au atoms were evaporated onto bulk MX_2, followed by the attachment of a thermal-release tape. The Au atoms make a bond with chalcogen atoms of the topmost layer of bulk MX_2. The bonding of Au atoms with the topmost layer of MX_2 is stronger than the interlayer-interaction MX_2, resulting in a selective peeling of the topmost layer as a single layer of TMD. The peeled-off layer can be stuck on the given substrate, and thermal-release tape can be removed by plasma etching. Finally, a monolayer TMD was obtained on the given substrate. This method can produce large-size flakes up to ~500 µm. The Au atom–mediated mechanical exfoliation gives large-size sheets than only tape-mediated exfoliation.

Monolayer TMD, including MoS_2, $NbSe_2$, and h-BN, were successfully prepared by mechanical exfoliation from bulk materials. This method can produce good-quality single-layer TMDs with low yield and smaller sizes, which is good for basic studies, but not for practical device applications. The bulk synthesis of TMD materials is possible by liquid-phase exfoliation. Coleman et al. reported 2D nanosheets by liquid-phase exfoliation of layered materials [79]. They started the sonication of commercially available MoS_2, WS_2, and BN powder in different solvents of varying surface tension. The resultant dispersions were centrifuged, followed by the separation of exfoliated materials. These dispersions of 2D TMD can be dispersed in common solvents, but they can be effectively dispersed in solvents of specific surface tension of ~40 mJ/m^2. Based on theoretical analysis, it was proposed that the solvents which minimize the energy of exfoliation can be effectively used for the exfoliation of 2D-layered materials. The studies suggested that various layered materials can be successfully exfoliated using N-methyl-pyrrolidone (NMP) and isopropanol (IPA) solvents. The exfoliated single-flake 2D materials can be deposited on the required substrate or form the films. Further, TEM analysis confirms 2D nanosheets structure of the liquid-phase exfoliated MoS_2, WS_2, and BN samples with lateral sizes ranging from 50 nm to 1,000 nm. The SEM and AFM characterization of liquid-phase exfoliated 2D materials is shown in Figures 3.8a. Figure 3.8b shows the proposed liquid-phase exfoliated mechanism of 2D materials [80].

The synthesis of TMDs nanosheets was also demonstrated by the CVD technique. Few-layer large-area MoS_2 nanosheets were grown by sulfurization of Mo metal film by sulfur vapor [81, 82]. The thickness and size of the MoS_2 film can be controlled by controlling the thickness of predeposited Mo film on the SiO_2 substrate. MoS_2 nanosheets can also be produced by using transition metal oxide and chlorides as Mo sources. Many ultrathin 2D TMDCs materials, including WS_2, $MoSe_2$, WSe_2,

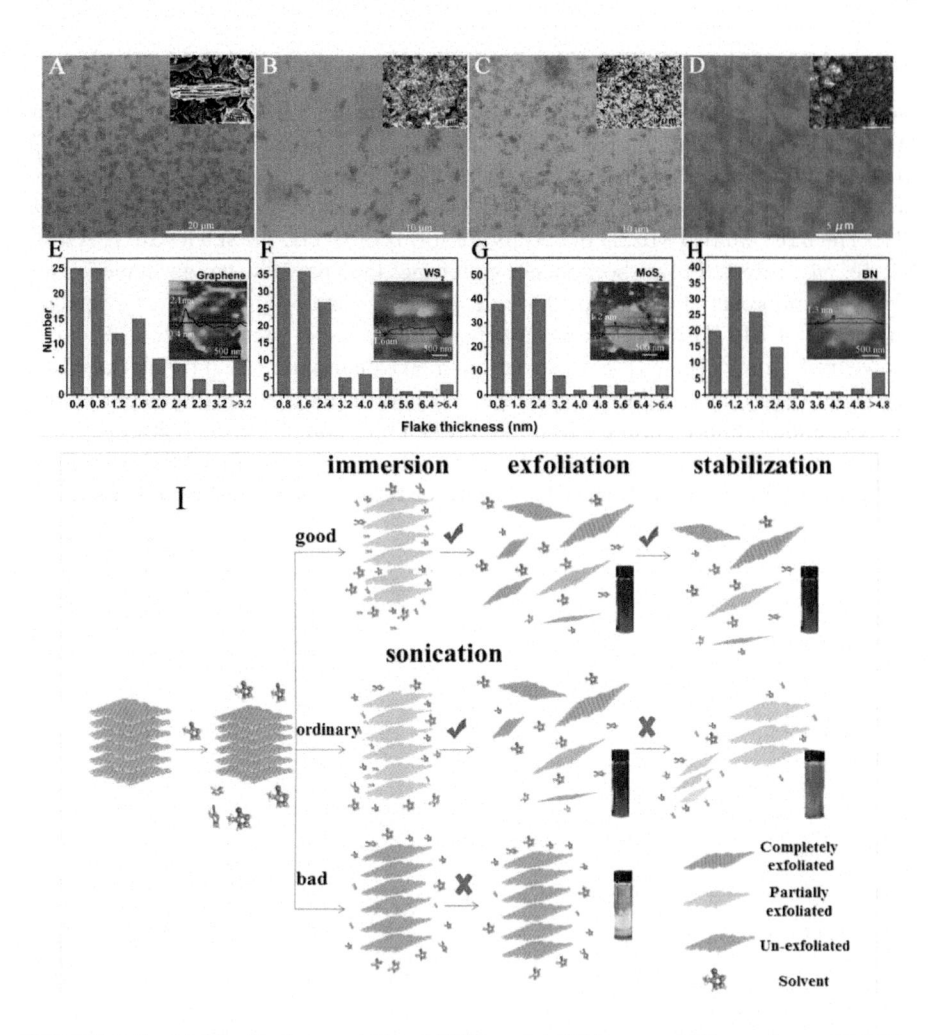

FIGURE 3.8 (a) Liquid-phase exfoliated SEM and AFM image of (A) graphene, (B) WS$_2$, (C) MoS$_2$, and (D) h-BN. The inset in each figure shows the corresponding original sample. (E–F) AFM image and histograms of the numbers of different flake thicknesses of graphene, WS$_2$, MoS$_2$, and h-BN, obtained after four weeks of centrifugation. (b) Shows the proposed liquid-phase exfoliation mechanism of 2D materials.

Source: Reproduced with permission from ref. [80].

ZrS$_2$, ReS$_2$, and MoTe$_2$, have been synthesized by the CVD technique under different experimental conditions [82]. The mechanical and electrical properties of TMDC materials have been evaluated. For example, the in-plane stiffness and Young's modulus of single-layer MoS$_2$ were reported to be 180 ± 60 Nm^{-1} and 270 ± 100 GPa, respectively [83]. TMD materials have shown metallic as well as semiconducting properties. The tunable electronic properties of TMD are due to the filling of the non-bonding d bands from group 4 to the group 10 elements.

3.4 MONOELEMENTAL GRAPHENE ANALOGOUS 2D MATERIALS

Atomically thin materials formed by a single element are referred to as monoelemental 2D materials. Similar to graphene, the family of monoelemental 2D (ME2D) materials has received much attention due to their high mobilities and semiconductor properties. ME2Ds form from group IIIA–VIA elements, as shown in Figure 3.9a [84]. The most studied ME2D materials are forms of B, Ga, In, Si, Ge, Sn, P, As, Bi, Se, Te, etc., referred to as borophene, gallenene, indiene, silicene, germanene, stanene, phosphorene, arsenene, bismuthine, selenene, and tellurene. After graphene, a timeline of ME2D materials development, either experimental or theoretical, is shown in Figure 3.9b [84]. Similar to other 2D materials, ME2D materials have been synthesized by top-down approaches, such as the mechanical/liquid exfoliation method, and bottom-up approaches, like CVD, PVD, and chemical synthesis.

After the discovery of graphene in 2004, silicene was the first ME2D material added to the family of 2D materials. Theoretical calculations predicted that silicene

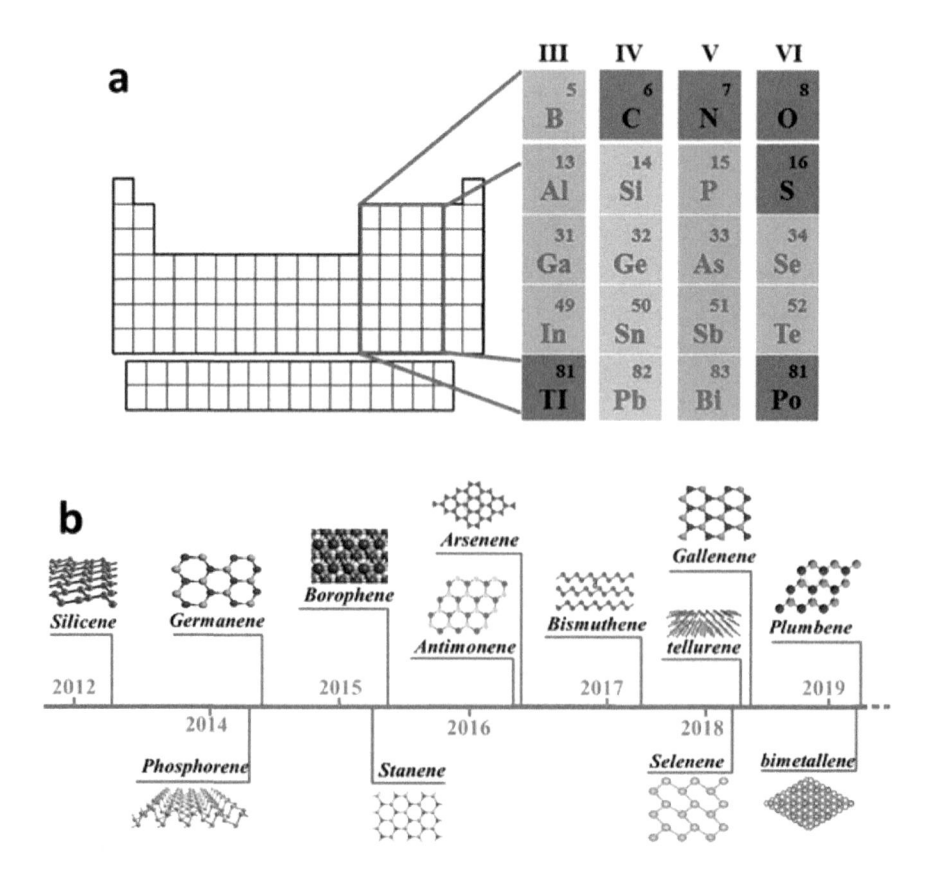

FIGURE 3.9 (a) Overview of graphene-analogous main-group 2D materials. (b) A timeline of progress of experimental realization of monoelemental 2D materials.

Source: Reproduced with permission from ref. [84].

has graphene-like properties, like massless Dirac fermions and quantum spin hall effect. Since then, silicene has been significantly studied for various applications. Silicon is well-known for its electronic applications, so silicene has the potential to be used in integrated electronics/circuits. Theoretical calculations predict a buckled honeycomb structure of monolayer silicene. Synthesis of freestanding monolayer silicene through the exfoliation method is difficult because of the lack of a layered structure. So silicene synthesis is demonstrated through the epitaxial growth on certain substrates under appropriate experimental conditions. Vogt et al. reported the first experimental study on silicene fabrication through in situ growth on Ag(111) substrate under ultrahigh vacuum [85]. Later studies found that multiple phases can exist in the epitaxial growth of silicene on Ag(111) substrate, which mainly depends on the substrate temperature and silicon flux. Silicene was fabricated on other substrates, including MoS_2, Ir(111), and $ZrB_2(0001)$, under an ultrahigh vacuum. Silicene on these substrates formed a monolayer honeycomb structure rather than multiple phases as obtained in the case of Ag(111) substrate. The study suggested that monolayer silicene shows metallic or semimetallic properties. However, monolayer silicene stability under ambient conditions is the main concern for its potential application.

The next ME2D material phosphorene is isolated from bulk phosphorous. Bulk phosphorous has many allotropes, including red, black, and white phosphorous. According to the literature, phosphorene is mainly isolated from black phosphorous, which has an orthorhombic crystal structure composed of weakly stacked layers. These stacked layers form corrugated rows, resulting in anisotropic mechanical, electrical, thermal, and optical properties of phosphorene [86]. Phosphorene can be synthesized using the micromechanical exfoliation method and CVD technique. The micromechanical process produces monolayer phosphorene, but it is difficult to control its thickness. The CVD method is most studied for the control fabrication of high-quality phosphorene. Significant research progress has been made on the thin films fabrication of black phosphorous directly on a particular substrate under an ultrahigh vacuum [87, 88]. Recently, monolayer blue phosphorous was grown on Au(111) under ultrahigh vacuum conditions, with a bandgap of 1.1 eV [89]. Due to the oxidized nature of phosphorene, it can be processed into deoxygenated aqueous solution. Therefore, phosphorene requires a unique passivation process for practical applications. The stability and scalability of phosphorene were evaluated by encapsulation methods under polymer or hexagonal boron nitride layers and thin aluminum oxide layers (AlOx). It was found that AlOx-encapsulated phosphorene shows long-term stability [90]. Further studies show that n-type or p-type doping is required for tuning charge carrier concentration in phosphorene. For example, diazonium salts form a stable passivation layer on black phosphorus, which results in p-type doping and enhancement of ambient stability. Adam et al. reported phosphorene *via* liquid-phase exfoliation of black phosphorus crystals (Figure 3.10) [91]. First, phosphorus crystals were sonicated in isopropanol for 16 h. At the time of sonication, the color of phosphorus was changed from black to reddish-brown to yellow (Figure 3.10c). The color change of suspension indicates the change in electronic structure of phosphene crystals. Further, SEM image (Figure 3.10d) showed the lateral size of phosphene flakes from 50 nm to 50 μm. Synthesis of other ME2D materials has been reviewed in recent review articles [86, 92, 93].

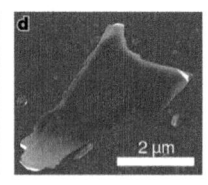

FIGURE 3.10   (a) Image of chemical vapor transport–grown black phosphorous, (b) illustration of monolayer phosphorene, (c) phosphorene suspension in isopropanol, and (d) SEM image of isopropanol-exfoliated phosphorene.

Source: Reproduced with permission from ref. [91].

REFERENCES

1. Novoselov KS, Geim AK, Morozov SV, Jiang D, Zhang Y, Dubonos SV, Grigorieva IV, Firsov AA. Electric field effect in atomically thin carbon films. Science. 2004;306:666–9.
2. Tan C, Cao X, Wu X-J, He Q, Yang J, Zhang X, Chen J, Zhao W, Han S, Nam G-H, Sindoro M, Zhang H. Recent advances in ultrathin two-dimensional nanomaterials. Chemical Reviews. 2017;117:6225–331.
3. Bian R, Li C, Liu Q, Cao G, Fu Q, Meng P, Zhou J, Liu F, Liu Z. Recent progress in the synthesis of novel two-dimensional van der Waals materials. National Science Review. 2021:nwab164.
4. Ferrari AC, Bonaccorso F, Fal'ko V, Novoselov KS, Roche S, Bøggild P, Borini S, Koppens FHL, Palermo V, Pugno N, Garrido JA, Sordan R, Bianco A, Ballerini L, Prato M, Lidorikis E, Kivioja J, Marinelli C, Ryhänen T, Morpurgo A, Coleman JN, Nicolosi V, Colombo L, Fert A, Garcia-Hernandez M, Bachtold A, Schneider GF, Guinea F, Dekker C, Barbone M, Sun Z, Galiotis C, Grigorenko AN, Konstantatos G, Kis A, Katsnelson M, Vandersypen L, Loiseau A, Morandi V, Neumaier D, Treossi E, Pellegrini V, Polini M, Tredicucci A, Williams GM, Hee Hong B, Ahn J-H, Min Kim J, Zirath H, van Wees BJ, van der Zant H, Occhipinti L, Di Matteo A, Kinloch IA, Seyller T, Quesnel E, Feng X, Teo K, Rupesinghe N, Hakonen P, Neil SRT, Tannock Q, Löfwander T, Kinaret J. Science and technology roadmap for graphene, related two-dimensional crystals, and hybrid systems. Nanoscale. 2015;7:4598–810.
5. Shekhirev M, Shuck CE, Sarycheva A, Gogotsi Y. Characterization of MXenes at every step, from their precursors to single flakes and assembled films. Progress in Materials Science. 2021;120:100757.
6. Zhang H. Ultrathin two-dimensional nanomaterials. ACS Nano. 2015;9:9451–69.
7. Neugebauer P, Orlita M, Faugeras C, Barra AL, Potemski M. How perfect can graphene be? Physical Review Letters. 2009;103:136403.
8. Mayorov AS, Elias DC, Mukhin IS, Morozov SV, Ponomarenko LA, Novoselov KS, Geim AK, Gorbachev RV. How close can one approach the dirac point in graphene experimentally? Nano Letters. 2012;12:4629–34.
9. Balandin AA, Ghosh S, Bao W, Calizo I, Teweldebrhan D, Miao F, Lau CN. Superior thermal conductivity of single-layer graphene. Nano Letters. 2008;8:902–7.
10. Lee C, Wei X, Kysar Jeffrey W, Hone J. Measurement of the elastic properties and intrinsic strength of monolayer graphene. Science. 2008;321:385–8.
11. Nair RR, Blake P, Grigorenko AN, Novoselov KS, Booth TJ, Stauber T, Peres NMR, Geim AK. Fine structure constant defines visual transparency of graphene. Science. 2008;320:1308.
12. Bonaccorso F, Lombardo A, Hasan T, Sun Z, Colombo L, Ferrari AC. Production and processing of graphene and 2d crystals. Materials Today. 2012;15:564–89.

13. Shukla A, Kumar R, Mazher J, Balan A. Graphene made easy: high quality, large-area samples. Solid State Communications. 2009;149:718–21.

14. Moldt T, Eckmann A, Klar P, Morozov SV, Zhukov AA, Novoselov KS, Casiraghi C. High-yield production and transfer of graphene flakes obtained by anodic bonding. ACS Nano. 2011;5:7700–6.

15. Dhar S, Barman AR, Ni GX, Wang X, Xu XF, Zheng Y, Tripathy S, Ariando RA, Loh KP, Rubhausen M, Neto AHC, Özyilmaz B, Venkatesan T. A new route to graphene layers by selective laser ablation. AIP Advances. 2011;1:022109.

16. Hernandez Y, Nicolosi V, Lotya M, Blighe FM, Sun Z, De S, McGovern IT, Holland B, Byrne M, Gun'Ko YK, Boland JJ, Niraj P, Duesberg G, Krishnamurthy S, Goodhue R, Hutchison J, Scardaci V, Ferrari AC, Coleman JN. High-yield production of graphene by liquid-phase exfoliation of graphite. Nature Nanotechnology. 2008;3:563–8.

17. Hasan T, Torrisi F, Sun Z, Popa D, Nicolosi V, Privitera G, Bonaccorso F, Ferrari AC. Solution-phase exfoliation of graphite for ultrafast photonics. physica status solidi (b). 2010;247:2953–7.

18. Kennedy GL, Jr., Sherman H. Acute and subchronic toxicity of dimethylformamide and dimethylacetamide following various routes of administration. Drug and Chemical Toxicology. 1986;9:147–70.

19. O'Neill A, Khan U, Nirmalraj PN, Boland J, Coleman JN. Graphene dispersion and exfoliation in low boiling point solvents. The Journal of Physical Chemistry C. 2011;115:5422–8.

20. Wang S, Zhang Y, Abidi N, Cabrales L. Wettability and surface free energy of graphene films. Langmuir. 2009;25:11078–81.

21. Lotya M, Hernandez Y, King PJ, Smith RJ, Nicolosi V, Karlsson LS, Blighe FM, De S, Wang Z, McGovern IT, Duesberg GS, Coleman JN. Liquid phase production of graphene by exfoliation of graphite in surfactant/water solutions. Journal of the American Chemical Society. 2009;131:3611–20.

22. Brodie BC XIII. On the atomic weight of graphite. Philosophical Transactions of the Royal Society of London. 1859;149:249–59.

23. Staudenmaier L. Verfahren zur Darstellung der Graphitsäure. Berichte der deutschen chemischen Gesellschaft. 1898;31:1481–7.

24. Hummers WS, Offeman RE. Preparation of graphitic oxide. Journal of the American Chemical Society. 1958;80:1339.

25. Kumar P, Maiti UN, Lee KE, Kim SO. Rheological properties of graphene oxide liquid crystal. Carbon. 2014;80:453–61.

26. Kumar P, Shahzad F, Yu S, Hong SM, Kim Y-H, Koo CM. Large-area reduced graphene oxide thin film with excellent thermal conductivity and electromagnetic interference shielding effectiveness. Carbon. 2015;94:494–500.

27. Dimiev AM, Tour JM. Mechanism of graphene oxide formation. ACS Nano. 2014;8:3060–8.

28. Behabtu N, Lomeda JR, Green MJ, Higginbotham AL, Sinitskii A, Kosynkin DV, Tsentalovich D, Parra-Vasquez ANG, Schmidt J, Kesselman E, Cohen Y, Talmon Y, Tour JM, Pasquali M. Spontaneous high-concentration dispersions and liquid crystals of graphene. Nature Nanotechnology. 2010;5:406–11.

29. Kim JE, Han TH, Lee SH, Kim JY, Ahn CW, Yun JM, Kim SO. Graphene oxide liquid crystals. Angewandte Chemie International Edition. 2011;50:3043–7.

30. Padmajan Sasikala S, Lim J, Kim IH, Jung HJ, Yun T, Han TH, Kim SO. Graphene oxide liquid crystals: a frontier 2D soft material for graphene-based functional materials. Chemical Society Reviews. 2018;47:6013–45.

31. Narayan R, Kim JE, Kim JY, Lee KE, Kim SO. Graphene oxide liquid crystals: discovery, evolution and applications. Advanced Materials. 2016;28:3045–68.

32. Kumar P, Yu S, Shahzad F, Hong SM, Kim Y-H, Koo CM. Ultrahigh electrically and thermally conductive self-aligned graphene/polymer composites using large-area reduced graphene oxides. Carbon. 2016;101:120–8.

33. Kumar P, Kumar A, Cho KY, Das TK, Sudarsan V. An asymmetric electrically conducting self-aligned graphene/polymer composite thin film for efficient electromagnetic interference shielding. AIP Advances. 2017;7:015103.

34. Cho KY, Yeom YS, Seo HY, Kumar P, Lee AS, Baek K-Y, Yoon HG. Molybdenum-doped PdPt@Pt core—shell octahedra supported by ionic block copolymer-functionalized graphene as a highly active and durable oxygen reduction electrocatalyst. ACS Applied Materials & Interfaces. 2017;9:1524–35.

35. Cho KY, Seo HY, Yeom YS, Kumar P, Lee AS, Baek K-Y, Yoon HG. Stable 2D-structured supports incorporating ionic block copolymer-wrapped carbon nanotubes with graphene oxide toward compact decoration of metal nanoparticles and high-performance nanocatalysis. Carbon. 2016;105:340–52.

36. Cho KY, Yeom YS, Seo HY, Kumar P, Lee AS, Baek K-Y, Yoon HG. Ionic block copolymer doped reduced graphene oxide supports with ultra-fine Pd nanoparticles: strategic realization of ultra-accelerated nanocatalysis. Journal of Materials Chemistry A. 2015;3:20471–6.

37. Cho KY, Yeom YS, Seo HY, Kumar P, Baek K-Y, Yoon HG. A facile synthetic route for highly durable mesoporous platinum thin film electrocatalysts based on graphene: morphological and support effects on the oxygen reduction reaction. Journal of Materials Chemistry A. 2017;5:3129–35.

38. Yadav MK, Panwar N, Singh S, Kumar P. Preheated self-aligned graphene oxide for enhanced room temperature hydrogen storage. International Journal of Hydrogen Energy. 2020;45:19561–6.

39. Shahzad F, Kumar P, Kim Y-H, Hong SM, Koo CM. Biomass-derived thermally annealed interconnected sulfur-doped graphene as a shield against electromagnetic interference. ACS Applied Materials & Interfaces. 2016;8:9361–9.

40. Coraux J, N'Diaye AT, Busse C, Michely T. Structural coherency of graphene on Ir(111). Nano Letters. 2008;8:565–70.

41. Yu Q, Lian J, Siriponglert S, Li H, Chen YP, Pei S-S. Graphene segregated on Ni surfaces and transferred to insulators. Applied Physics Letters. 2008;93:113103.

42. Li X, Cai W, Colombo L, Ruoff RS. Evolution of graphene growth on Ni and Cu by carbon isotope labeling. Nano Letters. 2009;9:4268–72.

43. Reina A, Jia X, Ho J, Nezich D, Son H, Bulovic V, Dresselhaus MS, Kong J. Large area, few-layer graphene films on arbitrary substrates by chemical vapor deposition. Nano Letters. 2009;9:30–5.

44. Li X, Cai W, An J, Kim S, Nah J, Yang D, Piner R, Velamakanni A, Jung I, Tutuc E, Banerjee Sanjay K, Colombo L, Ruoff Rodney S. Large-area synthesis of high-quality and uniform graphene films on copper foils. Science. 2009;324:1312–4.

45. Bae S, Kim H, Lee Y, Xu X, Park J-S, Zheng Y, Balakrishnan J, Lei T, Ri Kim H, Song YI, Kim Y-J, Kim KS, Özyilmaz B, Ahn J-H, Hong BH, Iijima S. Roll-to-roll production of 30-inch graphene films for transparent electrodes. Nature Nanotechnology. 2010;5:574–8.

46. Petrone N, Dean CR, Meric I, van der Zande AM, Huang PY, Wang L, Muller D, Shepard KL, Hone J. Chemical vapor deposition-derived graphene with electrical performance of exfoliated graphene. Nano Letters. 2012;12:2751–6.

47. Kim R-H, Bae M-H, Kim DG, Cheng H, Kim BH, Kim D-H, Li M, Wu J, Du F, Kim H-S, Kim S, Estrada D, Hong SW, Huang Y, Pop E, Rogers JA. Stretchable, transparent graphene interconnects for arrays of microscale inorganic light emitting diodes on rubber substrates. Nano Letters. 2011;11:3881–6.

48. Vlassiouk I, Fulvio P, Meyer H, Lavrik N, Dai S, Datskos P, Smirnov S. Large scale atmospheric pressure chemical vapor deposition of graphene. Carbon. 2013;54:58–67.
49. Terasawa T-O, Saiki K. Growth of graphene on Cu by plasma enhanced chemical vapor deposition. Carbon. 2012;50:869–74.
50. An CJ, Kim SJ, Choi HO, Kim DW, Jang SW, Jin ML, Park J-M, Choi JK, Jung H-T. Ultraclean transfer of CVD-grown graphene and its application to flexible organic photovoltaic cells. Journal of Materials Chemistry A. 2014;2:20474–80.
51. Othmen R, Arezki H, Ajlani H, Cavanna A, Boutchich M, Oueslati M, Madouri A. Direct transfer and Raman characterization of twisted graphene bilayer. Applied Physics Letters. 2015;106:103107.
52. Forbeaux I, Themlin JM, Charrier A, Thibaudau F, Debever JM. Solid-state graphitization mechanisms of silicon carbide 6H—SiC polar faces. Applied Surface Science. 2000;162–163:406–12.
53. de Heer Walt A, Berger C, Ruan M, Sprinkle M, Li X, Hu Y, Zhang B, Hankinson J, Conrad E. Large area and structured epitaxial graphene produced by confinement controlled sublimation of silicon carbide. Proceedings of the National Academy of Sciences. 2011;108:16900–5.
54. Tromp RM, Hannon JB. Thermodynamics and Kinetics of Graphene Growth on SiC(0001). Physical Review Letters. 2009;102:106104.
55. Mishra N, Boeckl J, Motta N, Iacopi F. Graphene growth on silicon carbide: a review. Physica Status Solidi (A). 2016;213:2277–89.
56. Lin YM, Dimitrakopoulos C, Jenkins KA, Farmer DB, Chiu HY, Grill A, Avouris P. 100-GHz transistors from wafer-scale epitaxial graphene. Science. 2010;327:662.
57. Higginbotham AL, Kosynkin DV, Sinitskii A, Sun Z, Tour JM. Lower-defect graphene oxide nanoribbons from multiwalled carbon nanotubes. ACS Nano. 2010;4:2059–69.
58. Lim J, Narayan Maiti U, Kim N-Y, Narayan R, Jun Lee W, Sung Choi D, Oh Y, Min Lee J, Yong Lee G, Hun Kang S, Kim H, Kim Y-H, Ouk Kim S. Dopant-specific unzipping of carbon nanotubes for intact crystalline graphene nanostructures. Nature Communications. 2016;7:10364.
59. Naguib M, Kurtoglu M, Presser V, Lu J, Niu J, Heon M, Hultman L, Gogotsi Y, Barsoum MW. Two-dimensional nanocrystals produced by exfoliation of Ti3AlC2. Advanced Materials. 2011;23:4248–53.
60. Deysher G, Shuck CE, Hantanasirisakul K, Frey NC, Foucher AC, Maleski K, Sarycheva A, Shenoy VB, Stach EA, Anasori B, Gogotsi Y. Synthesis of Mo4VAlC4 MAX phase and two-dimensional Mo4VC4 MXene with five atomic layers of transition metals. ACS Nano. 2020;14:204–17.
61. Kumar P, Singh S, Hashmi SAR, Kim K-H. MXenes: emerging 2D materials for hydrogen storage. Nano Energy. 2021;85:105989.
62. Dwivedi N, Dhand C, Kumar P, Srivastava AK. Emergent 2D materials for combating infectious diseases: the potential of MXenes and MXene—graphene composites to fight against pandemics. Materials Advances. 2021;2:2892–905.
63. Alhabeb M, Maleski K, Anasori B, Lelyukh P, Clark L, Sin S, Gogotsi Y. Guidelines for synthesis and processing of two-dimensional titanium carbide (Ti3C2Tx MXene). Chemistry of Materials. 2017;29:7633–44.
64. Halim J, Lukatskaya MR, Cook KM, Lu J, Smith CR, Näslund L-Å, May SJ, Hultman L, Gogotsi Y, Eklund P, Barsoum MW. Transparent conductive two-dimensional titanium carbide epitaxial thin films. Chemistry of Materials. 2014;26:2374–81.
65. Wang X, Garnero C, Rochard G, Magne D, Morisset S, Hurand S, Chartier P, Rousseau J, Cabioc'h T, Coutanceau C, Mauchamp V, Célérier S. A new etching environment (FeF3/HCl) for the synthesis of two-dimensional titanium carbide MXenes: a route towards selective reactivity vs. water. Journal of Materials Chemistry A. 2017;5:22012–23.

66. Ghidiu M, Lukatskaya MR, Zhao M-Q, Gogotsi Y, Barsoum MW. Conductive two-dimensional titanium carbide 'clay' with high volumetric capacitance. Nature. 2014;516:78–81.
67. Naguib M, Mochalin VN, Barsoum MW, Gogotsi Y. 25th anniversary article: MXenes: a new family of two-dimensional materials. Advanced Materials. 2014;26:992–1005.
68. Li T, Yao L, Liu Q, Gu J, Luo R, Li J, Yan X, Wang W, Liu P, Chen B, Zhang W, Abbas W, Naz R, Zhang D. Fluorine-free synthesis of high-purity Ti3C2Tx (T=OH, O) via alkali treatment. Angewandte Chemie International Edition. 2018;57:6115–9.
69. Xuan J, Wang Z, Chen Y, Liang D, Cheng L, Yang X, Liu Z, Ma R, Sasaki T, Geng F. Organic-base-driven intercalation and delamination for the production of functionalized titanium carbide nanosheets with superior photothermal therapeutic performance. Angewandte Chemie International Edition. 2016;55:14569–74.
70. Sun W, Shah SA, Chen Y, Tan Z, Gao H, Habib T, Radovic M, Green MJ. Electrochemical etching of Ti2AlC to Ti2CTx (MXene) in low-concentration hydrochloric acid solution. Journal of Materials Chemistry A. 2017;5:21663–8.
71. Pang S-Y, Wong Y-T, Yuan S, Liu Y, Tsang M-K, Yang Z, Huang H, Wong W-T, Hao J. Universal strategy for HF-free facile and rapid synthesis of two-dimensional MXenes as multifunctional energy materials. Journal of the American Chemical Society. 2019;141:9610–6.
72. Lipatov A, Lu H, Alhabeb M, Anasori B, Gruverman A, Gogotsi Y, Sinitskii A. Elastic properties of 2D Ti3C2Tx MXene monolayers and bilayers. Science Advances. 2018;4:eaat0491.
73. Lipatov A, Alhabeb M, Lu H, Zhao S, Loes MJ, Vorobeva NS, Dall'Agnese Y, Gao Y, Gruverman A, Gogotsi Y, Sinitskii A. Electrical and elastic properties of individual single-layer Nb4C3Tx MXene flakes. Advanced Electronic Materials. 2020;6:1901382.
74. Li G, Kushnir K, Dong Y, Chertopalov S, Rao AM, Mochalin VN, Podila R, Titova LV. Equilibrium and non-equilibrium free carrier dynamics in 2D Ti_3C_2Tx MXenes: THz spectroscopy study. 2D Materials. 2018;5:035043.
75. Li G, Natu V, Shi T, Barsoum MW, Titova LV. Two-Dimensional MXenes Mo2Ti2C3Tz and Mo2TiC2Tz: microscopic conductivity and dynamics of photoexcited carriers. ACS Applied Energy Materials. 2020;3:1530–9.
76. Han M, Shuck CE, Rakhmanov R, Parchment D, Anasori B, Koo CM, Friedman G, Gogotsi Y. Beyond Ti3C2Tx: MXenes for electromagnetic interference shielding. ACS Nano. 2020;14:5008–16.
77. Chhowalla M, Shin HS, Eda G, Li L-J, Loh KP, Zhang H. The chemistry of two-dimensional layered transition metal dichalcogenide nanosheets. Nature Chemistry. 2013;5:263–75.
78. Desai SB, Madhvapathy SR, Amani M, Kiriya D, Hettick M, Tosun M, Zhou Y, Dubey M, Ager Iii JW, Chrzan D, Javey A. Gold-mediated exfoliation of ultralarge optoelectronically-perfect monolayers. Advanced Materials. 2016;28:4053–8.
79. Coleman Jonathan N, Lotya M, O'Neill A, Bergin Shane D, King Paul J, Khan U, Young K, Gaucher A, De S, Smith Ronan J, Shvets Igor V, Arora Sunil K, Stanton G, Kim H-Y, Lee K, Kim Gyu T, Duesberg Georg S, Hallam T, Boland John J, Wang Jing J, Donegan John F, Grunlan Jaime C, Moriarty G, Shmeliov A, Nicholls Rebecca J, Perkins James M, Grieveson Eleanor M, Theuwissen K, McComb David W, Nellist Peter D, Nicolosi V. Two-dimensional nanosheets produced by liquid exfoliation of layered materials. Science. 2011;331:568–71.
80. Shen J, He Y, Wu J, Gao C, Keyshar K, Zhang X, Yang Y, Ye M, Vajtai R, Lou J, Ajayan PM. Liquid phase exfoliation of two-dimensional materials by directly probing and matching surface tension components. Nano Letters. 2015;15:5449–54.
81. Lee Y-H, Zhang X-Q, Zhang W, Chang M-T, Lin C-T, Chang K-D, Yu Y-C, Wang JT-W, Chang C-S, Li L-J, Lin T-W. Synthesis of large-area MoS2 atomic layers with chemical vapor deposition. Advanced Materials. 2012;24:2320–5.

82. Zhan Y, Liu Z, Najmaei S, Ajayan PM, Lou J. Large-area vapor-phase growth and characterization of MoS2 atomic layers on a SiO2 substrate. Small. 2012;8:966–71.
83. Bertolazzi S, Brivio J, Kis A. Stretching and breaking of ultrathin MoS2. ACS Nano. 2011;5:9703–9.
84. Yang F, Song P, Xu W. The applications of 2D nanomaterials in energy-related process. Adapting 2D nanomaterials for advanced applications. ACS Symposium Series. 2020;1353:219–51.
85. Vogt P, De Padova P, Quaresima C, Avila J, Frantzeskakis E, Asensio MC, Resta A, Ealet B, Le Lay G. Silicene: compelling experimental evidence for graphenelike two-dimensional silicon. Physical Review Letters. 2012;108:155501.
86. Mannix AJ, Kiraly B, Hersam MC, Guisinger NP. Synthesis and chemistry of elemental 2D materials. Nature Reviews Chemistry. 2017;1:0014.
87. Smith JB, Hagaman D, Ji H-F. Growth of 2D black phosphorus film from chemical vapor deposition. Nanotechnology. 2016;27:215602.
88. Li X, Deng B, Wang X, Chen S, Vaisman M, Karato S-I, Pan G, Larry Lee M, Cha J, Wang H, Xia F. Synthesis of thin-film black phosphorus on a flexible substrate. 2D Materials. 2015;2:031002.
89. Zhang JL, Zhao S, Han C, Wang Z, Zhong S, Sun S, Guo R, Zhou X, Gu CD, Yuan KD, Li Z, Chen W. Epitaxial growth of single layer blue phosphorus: a new phase of two-dimensional phosphorus. Nano Letters. 2016;16:4903–8.
90. Wood JD, Wells SA, Jariwala D, Chen K-S, Cho E, Sangwan VK, Liu X, Lauhon LJ, Marks TJ, Hersam MC. Effective passivation of exfoliated black phosphorus transistors against ambient degradation. Nano Letters. 2014;14:6964–70.
91. Woomer AH, Farnsworth TW, Hu J, Wells RA, Donley CL, Warren SC. Phosphorene: synthesis, scale-up, and quantitative optical spectroscopy. ACS Nano. 2015;9:8869–84.
92. Zhou D, Li H, Si N, Li H, Fuchs H, Niu T. Epitaxial growth of main group monoelemental 2D materials. Advanced Functional Materials. 2021;31:2006997.
93. Zavabeti A, Jannat A, Zhong L, Haidry AA, Yao Z, Ou JZ. Two-dimensional materials in large-areas: synthesis, properties and applications. Nano-Micro Letters. 2020;12:66.

4 2D Materials as Antiviral Agents to Combat COVID-19

Chetna Dhand and Jamana Prasad Chaurasia

Throughout human history, microbial infections and infectious diseases have been the biggest killer [1]. Multiple-dose antibiotic administration is the most widely used treatment strategy to prevent infection in critical health problems, such as deep burns, diabetic ulcers, and other infections, and to treat existing infections. Overuse of antibiotics is partly to blame for the growth of multidrug-resistant illnesses, including superbugs, which seem to be the biggest threat to public health worldwide. Antibiotic-resistant microorganisms are responsible for 1 million fatalities per year and are resistant to the majority of medications now available. If suitable efforts are not implemented, this number is expected to rise to 10 million by 2050 [1, 2]. As a result, healthcare professionals are attempting to create novel antimicrobial medications to counteract the spread of bacterial resistance. Graphene oxide (GO), MoS_2, MXenes, and other 2D materials have been studied for their antibacterial properties throughout the previous decade, and numerous modes of action have been discovered. The rupturing of the phospholipid bilayer, enhanced permeability across the membrane, DNA degradation, decreased metabolism rate, and stress induced on the cell membrane have all been proven to have good bactericidal effects against both gram-positive and gram-negative bacteria [3–5]. 2D materials are emerging as a promising candidate to inhibit bacterial and fungal growth due to their high surface area, richness in chemical functionalities that allow for chemical alterations and functionalization, and enrichment with nucleophilic moieties, such as OH, F, or O, that can mediate the attachment of antimicrobial metallic adjuvants via coordination chemistry.

Researchers have been paying significant attention to 2D materials in recent years due to their wide variety of functional features, including huge surface area, high electrical mobility and conductivity, good electrochemical, mechanical, and piezoelectric properties, and antimicrobial/antiviral characteristics [6–16]. Thus, the multifunctional properties of 2D materials may lead to their use in the creation of a wide range of systems/devices that are also critical in the battle against infectious illnesses, such as COVID-19. Because they tend to disintegrate and kill bacteria and viruses, 2D materials, particularly metal ions–ornamented 2D materials, might be utilized to build antiviral and antimicrobial materials and surfaces for use in medical settings, high-touch surfaces, and a range of consumer products. In this chapter, we discuss the antiviral potential of 2D materials, including graphene-based materials

DOI: 10.1201/9781003316381-4

and MXene. We also propose prospects of 2D materials for antiviral applications and advocate how they can be implemented to control emerging pandemic threats.

4.1 GRAPHENE-BASED ANTIVIRAL SURFACES AND COATINGS

A novel lethal SAR-CoV-2 virus started spreading across people in December 2019 [17]. The most prevalent route for COVID-19 spread is by sub-micron-sized airborne droplets [18]. Furthermore, an individual can get the virus by touching their mouth, nose, or eyes after making contact with contaminated surfaces and equipment. SAR-CoV-2 viral life span varies on various surfaces, according to recent scientific studies [19]. When compared to copper surfaces (4 h), SARS-CoV-2 had a longer survival duration on plastic (72 h), cardboard (48 h), and stainless steel (24 h). Furthermore, the virus is more stable on smooth surfaces than on uneven surfaces, like tissue sheets (3 h), textiles (2 h), and wood (2 h). Surprisingly, even on day 7 [20], a measurable quantity of the virus has been observed on the exterior layer of surgical face masks. As a result, COVID-19 transmission can be aided by infected high-touch surfaces with high viral stability. The creation of effective preventive surfaces/surface coatings against COVID-19 can play a critical role in regulating viral propagation via high-contact components, goods, and systems in the ongoing pandemic scenario, when COVID-19 occurrences are dramatically growing each day globally.

The antimicrobial potential of graphene-related materials has been studied intensively [16, 21]. According to published research, GO and related materials have broad-spectrum inhibitory effects against gram-positive and gram-negative bacteria [22], yeast, and fungi [23]. Sametband et al. [24] presented the very first study demonstrating the antiviral effects of GO and partly reduced sulfonated GO (SO_3-rGO) against herpes simplex virus type 1 (HSV-1) via a competitive repressing mechanism in 2014. GO and SO_3-rGO have several negatively charged moieties, analogous to the cell surface receptor heparan sulphate, and hence, both groups compete for HSV-1 attachment. The key inhibitory factor in protecting Vero cells from invasion is blocking viral binding sites with nanomaterial. Ye *et al.* [25] examined the antiviral efficacy of precursor's graphite (Gt) and graphite oxide (GtO), GO, rGO, composite of GO with polyvinylpyrrolidone (PVP) and poly(diallyl dimethylammonium chloride) (PDDA) composite, and GO-poly(diallyl dimethylammonium chloride) (PDDA). The researchers discovered that GO has wide antiviral efficacy against both DNA virus, that is, pseudorabies virus (PRV), and RNA virus, that is, porcine epidemic diarrhea virus (PEDV). The antiviral capabilities of GO may possibly be due to its negatively charged, sharp-edged structure, according to the findings. Although GO conjugated with non-ionic polyvinylpyrrolidone polymer demonstrated substantial antiviral activity, GO coupled with cationic PDDA exhibited little virus suppression, suggesting that antiviral activity necessitates a negative charge. Ziem et al. developed thermally polysulfated dendritic polyglycerol functionalized rGO sheets. This rGO-based composite material's antiviral activity was established against a number of viruses, including orthopoxviruses, HSV1, and horse herpesvirus type 1 (EHV1) [26, 27].

Song et al. [28] developed a GO-based, label-free technique for identifying and sanitizing ecological viruses, including enterovirus 71 (EV71) and endemic gastrointestinal avian influenza A virus (H9N2), that are resistant to organic sanitizers and soaps and have a high level of environmental stability. Physiochemical interactions (hydrogen bonding interactions, electrostatic forces, and electron transfer reactions) between GO and viruses play a critical part in the capture and killing of viruses, according to the report. At elevated temperature circumstances, the virucidal capabilities of GO are observed to be improved (56°C). Chen et al. investigated the virucidal potential of GO and its nanocomposite with AgNPs against FCoV and IBDV [28]. Figure 4.1 depicts the schematic representation of these viruses and graphene-derived antiviral agents. GO-AgNPs were demonstrated to have a greater degree of virucidal activity compared to pure graphene oxide. As a result, nanocomposites of AgNPs with GO can limit FCoV by 25% and IBDV by 23%, while GO exhibited a suppression of 16% for FCoV virus and absolutely no inhibition for IBDV. The antiviral mechanism behind the antiviral efficiency of GO and GO-AgNPs nanocomposite may be elucidated using the structural properties shown in Figure 4.3. The antiviral characteristics of graphene-derived materials can be explained by their molecular interactions with viruses. The opposing dielectric characteristics of antiviral agents (GO and GO-AgNPs) over one hand, and both viruses on another, may cause graphene materials and viruses to bind together, according to this theory. GO may interact with the virus's membrane lipids in these conditions. The association of the silver and thiol groups of

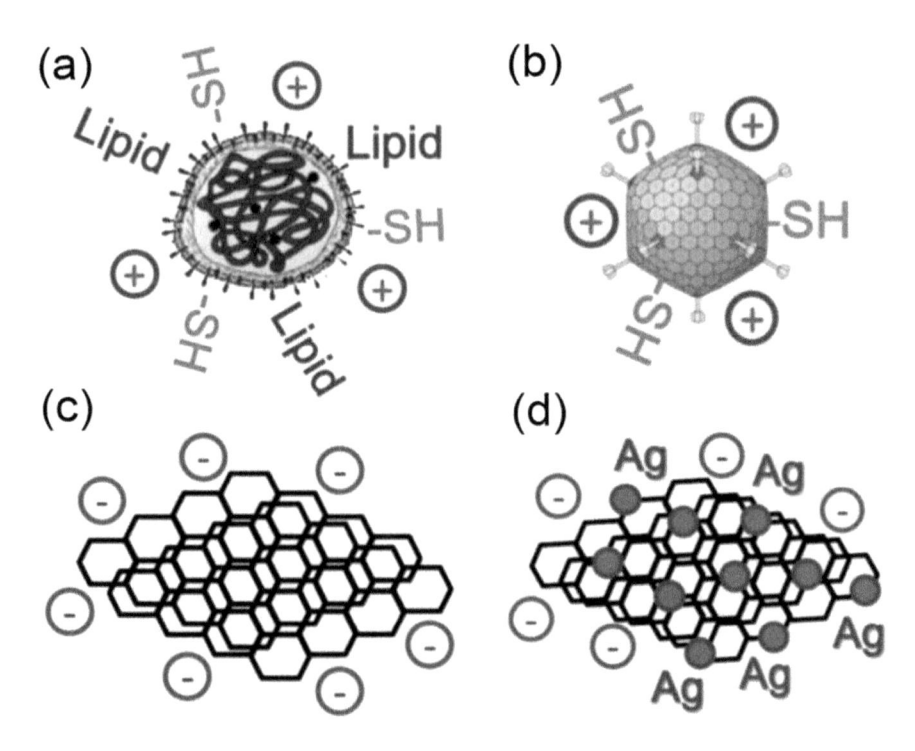

FIGURE 4.1   Schematic illustration of (a) the enveloped FCoV, (b) non-enveloped IBDV, (c) GO, and (d) GO-AgNPs nanocomposite.

Source: Reproduced from ref. [28].

virus proteins provides supplementary interfacial contacts in both enveloped and non-enveloped viruses. Controlling IBDV infection requires this binding. When it comes to GO-AgNPs and nonenveloped viruses, GO nanosheets have a minimal role in the overall effectiveness of the nanocomposite to kill viruses. By providing a platform for AgNPs to distribute uniformly and without aggregation, GO improves the antiviral agent's overall efficacy. Subsequent investigation verifies that the electrostatic interactions among graphene oxide/reduced graphene oxide and lipid membranes result in graphene adhesion to the viral surface and liposome breakdown [29].

The experiments mentioned earlier demonstrate the importance of accessible functional groups on the surface of the graphene-derived materials for effective binding to virus species. Recent simulation studies [30] have proven graphene's binding capabilities in its pristine state. Raval et al. show in simulations that virgin multilayered graphene generated by exfoliation of graphite mechanically has a high affinity for the spike receptor-binding region of the SARS-CoV-2 [30]. The binding energy involving graphene flake and the viruses grew with the increase in the number of graphene layers in this circumstance. For the binding process of seven-layer graphene flakes with the virus, the change in Gibb's free energy is estimated to be –28.01 Kcal/mol.

Carbon quantum dots (CQDs) feature a graphitic sp^2 hybridized structure and are less than 10 nm in size. To produce antiviral characteristics, CQDs may be readily functionalized with different functional groups. In research published by Aleksandra et al., boric acid functionalization of CQDs was done, and their antiviral potential was unveiled against HCoV-229E [31]. In this study, two modes of antiviral action were reported: (1) the binding of the CQDs with diameter of ~7 nm to the S protein of the virus that inhibits the interfacial interactions between the host cell and the virus required to initiate the infection, and (2) the involvement of CQDs in hindering the virus RNA replication. The existence of boronic acid functionalities was shown to be critical in determining antiviral efficacy. Barras et al. revealed consistent patterns where they revealed the antiviral activity of carbon nanodots functionalized with the 4-aminophenyl boronic acid hydrochloride against HSV1 virus [32]. The antiviral action of CQDs is highlighted in Figure 4.2.

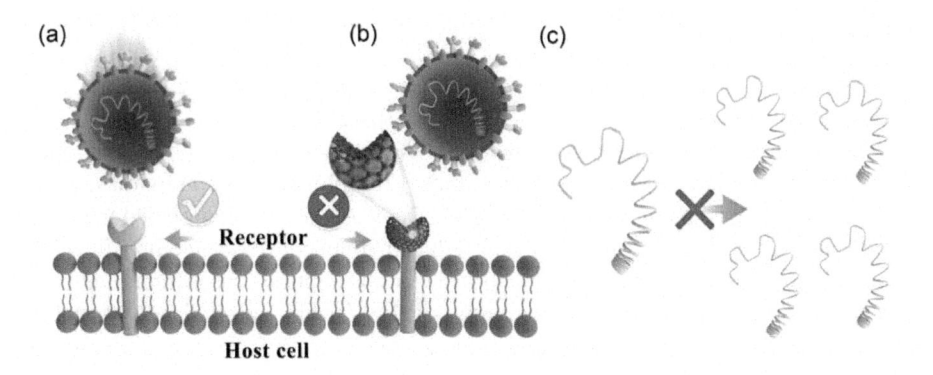

FIGURE 4.2   Schematic representation showing the antiviral mechanism of carbon quantum dots.

Source: Reproduced from ref. [33].

Yang et al. [34] created a multifunctional β-CD functionalized sulfonated graphene composite loaded with curcumin (GSCC) and evaluated its antiviral effectiveness against respiratory syncytial virus (RSV). RSV, like SARS-CoV-2, is a negative-sense virus that infects both the upper and lower respiratory tracts, mostly affecting children and the elderly. The findings showed that GSCC might prevent RSV from invading healthy cells by efficiently suppressing the viruses and blocking viral adhesion, as well as having prophylactic and curative actions against the virus. The antiviral impact of GO-AgNPs composite on the replication of porcine reproductive and respiratory syndrome virus (PRRSV) was examined in a recent study [35]. According to the findings, viral exposure to composites of GO-AgNPs prevents the virus from entering healthy cells with a 59.2% effectiveness and stimulates the synthesis of IFN-α stimulating genes and interferon-α that suppress the multiplication of the virion particles. Elechiguerra examined the antiviral properties of AgNPs coated with carbon, BSA, and PVP [36]. In terms of antiviral efficacy, carbon-coated silver nanoparticles outperformed the other samples, which might be owing to their surface characteristics, which increase virus receptivity.

GO and its derivatives have broad-spectrum antiviral effectiveness, that is, against both DNA and RNA viruses, against negative- and positive-sense viruses, against enclosed and non-enveloped viruses, and so on. These materials are rapidly progressing in the development of antiviral materials and coatings to protect surfaces from viral contamination and to prevent viral propagation and transmission through surfaces due to their virucidal activity. GO and related materials have been extensively explored as surface-coating materials to avoid contamination by the aggressive and contagious SARS-CoV-2 virus in the current COVID-19 pandemic scenario. GO and rGO-SO$_3$ nanocoatings augmented with copper ions and copper nanoparticles could be a promising material for the design and development of antiviral coatings against SARS-CoV-2 virus because the virus's structure is rich in -COOH functional groups and SARS-CoV-2 virus has the shortest survival time on copper surfaces. Nanocomposites of GO and rGO-SO$_3$ with antimicrobial metal ions and metal nanoparticles, such as Ti, Ag, and Au, should also be studied for the manufacture of extremely effective antiviral nanocoating that can prevent bacterial and viral spread via contaminated surfaces. These nanomaterials can assist in trapping and disrupting viral structures as well as reducing virus survival time on a variety of coated surfaces.

The integration of magnetic elements in the form of nanoparticles and other forms has been discovered to favorably impact the photothermal antiviral activity of graphene-derived materials. In this regard, Deokar et al. have fabricated the magnetic nanoparticles–enriched sulfonated-reduced graphene oxide composite [37]. This nanocomposite material was able to successfully capture near-infrared light and use the absorbed photon to photothermally kill HSV1 with an efficiency of around 99.99%, which was superior than MNPs alone. The high surface-to-volume ratio of the developed composite material, good photothermal characteristics of graphene, and strong intermolecular interactions between the magnetic nanoparticles and the virus surface are all contributors in this nanocomposite's outstanding performance.

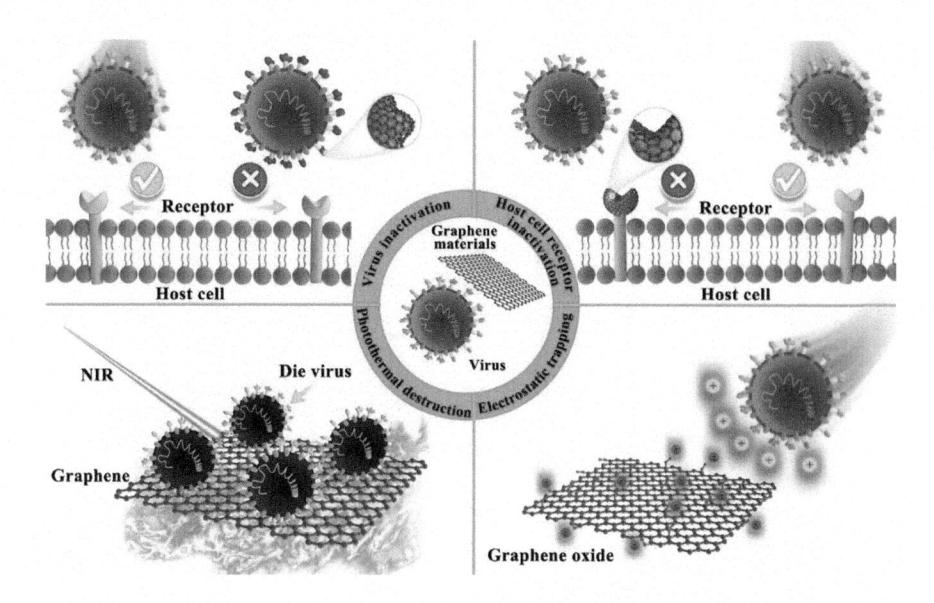

FIGURE 4.3 Schematic illustrations of the modes of antiviral action for graphene-derived materials.

Source: Reproduced from ref. [33].

Figure 4.3 depicts the key antiviral mechanisms for graphene and related materials enlightened by Seifi and Kamali et al. [33]. Table 4.1 highlights the antiviral capabilities of several graphene oxide derivatives as well as their mode of action.

4.2 MXENES-BASED ANTIVIRAL MATERIALS

MXene is a rapidly expanding family of 2D materials that have a wide range of chemistries and topologies [41–44]. MXenes are being used in biological applications for the first time, including photothermal treatment for tumors, theragnostic, sensors/biosensors, dialysis, and brain electrodes. Although MXene has previously been investigated for its antibacterial potential [45–47], relatively few publications on its antiviral activities are known. Recently, Unal et al. have investigated and compared the antiviral potential of four different MXenes, $Ti_3C_2T_x$, $Ta_4C_3T_x$, $Mo_2Ti_2C_3T_x$, and $Nb_4C_3T_x$, and performed detailed immune profiling experimentation [48]. To test this hypothesis, viral infection has been performed with distinct viral clades of the SARS-CoV-2 virus *in vitro* in the presence of various MXenes. Result outcome suggested good antiviral activity of Ti_3C_2Tx and $Mo_2Ti_2C_3T_x$ even at low concentrations. The interaction between the negatively charged, polar, electrochemically active surface of MXene with the virus proteins is reported to be responsible for viral inhibition. Further, antiviral efficacy differed among MXenes with comparable structures but varying thicknesses of surface atoms, showing the importance of surface chemistry. In addition to this, the report revealed that the antiviral effectiveness of the

TABLE 4.1

Various Graphene and Related Materials with Their Antiviral Efficiency against Different Viral Strains

Graphene-Derived Materials	Targeted Virus	Mechanism of Antiviral Action	Key Outcome of the Study	Ref.
Graphene oxide (GO) and reduced sulphonated graphene oxide	**Herpes simplex virus type I:** It is a DNA virus that infects 70–90% of the world population, predominantly affecting epithelial cells and neuron in the peripheral nervous system, causing oral lesions, ophthalmic disorders, and in rare cases, encephalitis.	Competitive inhibition mechanism.	Both the graphene-based materials showed virucidal activity at very low concentrations (ng/mL), resulting in the inhibition of plaque formation. These materials are found to be highly biocompatible.	[24]
GO, reduced graphene oxide (rGO), graphite (Gt), graphite oxide (GtO), GO-polyvinylpyrrolidone, GO-PDDA (poly (diallyl dimethylammonium chloride))	**Pseudorabies virus (PRV):** It's a DNA virus that mostly affects animals and causes problems with the central nervous system, respiratory system, and reproductive system. **Porcine epidemic diarrhea virus (PEDV):** It is an RNA virus that belongs to the alpha-coronavirus family. Pigs have a significant mortality rate due to this virus.	The negative charge available of the graphene sheets and its nanosheet morphology is unveiled to be responsible for its antiviral properties.	By breaking the virus's structure, GO inhibits both viruses (PRV, PEDV) during their early-entry phase into the virus. Unfortunately, once an infection has commenced, materials will not be able to stop it. Even at low non-cytotoxic concentrations (1.5 µg/mL), GO has antiviral effects. Nonionic GO-PVP displays antiviral activity similar to GO, indicating that the negative charge is essential for antiviral activities. Cationic GO-PDDA showed no antiviral effect. Multilaminate GtO had a considerably lesser inhibitory impact than single-layered GO and rGO, but Gt, which has a non-nanosheet structure, had no antiviral action, suggesting that the nanosheet structure is important. Antiviral action is dependent on the structure.	[25]

GO	**Enterovirus-71 (EV71):** It is an RNA-enveloped virus. It is responsible for hand-foot-and-mouth disorder, which is extremely common in children. **Endemic gastrointestinal avian influenza A virus (H9N2):** It is an enveloped RNA virus that causes bird flu and influenza commonly in humans.	The interactions among the GO and the virus result in destruction of the virus membrane and leakage of its genomic content.	GO strongly promotes absolute annihilation, exclusion, and dis-infection characteristics that help in achieving 6-log decrease in the infectivity by the virus. The GO's sterilizing impact against the virus is observed to be strengthened at higher temperatures and with longer exposure duration.	[38]
GO and GO-silver nanoparticles (AgNPs)	**Feline coronavirus (FCoV):** It is a positive-sense capsid RNA virus. This virus results in feline infectious peritonitis disease in cats, which is highly lethal. **Infectious bursal disease virus (IBDV):** It is an unenveloped RNA virus. In chickens, it is accountable for immunosuppression.	Strong intermolecular interactions among the GO and GO-AgNPs composite with the virus membrane that includes electrostatic interactions, van der Waals forces, interactions among AgNPs and -SH groups of membrane proteins, leading to the disintegration of virus structure.	The incorporation of Ag particles in GO sheets improves GO's viricidal activities against enclosed viruses while also widening its inhibitory impact on non-enveloped viruses.	[28]
Curcumin-loaded β-cyclodextrin functionalized sulfonated graphene	**Respiratory syncytial virus (RSV):** It is an enclosed RNA virus with a 150 nm average diameter. This virus causes respiratory problems in babies, children, the elderly, and individuals who are immunocompromised. Both the lower and upper respiratory tracts are affected.	This smart nanocomposite inhibits the virus by (i) interfering with the association/attachment of the virus with the host cell, (ii) by inhibiting the virus replication, and (iii) by destroying the virus membrane by its direction interactions with the virus.	The host cells are proven to be completely safe at the composite effective antiviral concentration. The viral titer shows a four-order drop in virus concentration after incubation with this smart nanocomposite.	[34]

(*Continued*)

TABLE 4.1 (CONTINUED)

GO-AgNPs	**Porcine reproductive and respiratory syndrome virus (PRRSV):** It is a positive-sense RNA virus that regularly affects the pigs.	The antiviral properties of the GO-AgNPs nanocomposite stem from its capacity to prevent virus entrance into the host cell while also boosting the synthesis of IFN-α-stimulating genes and interferon-α.	Plaque experiments reveal that the nanocomposites limit viral entry into the host cells, resulting in virus replication suppression.	[35]
Cationic curcumin functionalized carbon dots (CCM-CDs)	**PEDV.**	The antiviral activity of CCM-CDs is owing to the inhibition of virus entry, synthesis of RNA, budding of virus, and ROS accumulation.	The virus entry is inhibited by over 50%.	[39]
Sulphonated magnetic nanoparticles functionalized with reduced graphene oxide (SMRGO)	**HSV.**	The nanocomposite shows excellent photothermal inhibition of the virus.	No irradiation: 34.5% killing for MNPs and 35% killing for SMRGO were recorded. Upon NIR irradiation, 79% inhibition for MNPs and 99.99% killing for SMRGO are observed.	[37]
Cu-Gr nanocomposite	**Influenza A:** Similar to coronavirus, it is an enveloped RNA virus that affects the respiratory system and spreads through droplets.	Cu-Gr nanocomposite can significantly reduce viral infectivity, as seen by the decreased viral entrance, gene expression, and subsequent formation of progeny virions.	50% viral reduction.	[40]

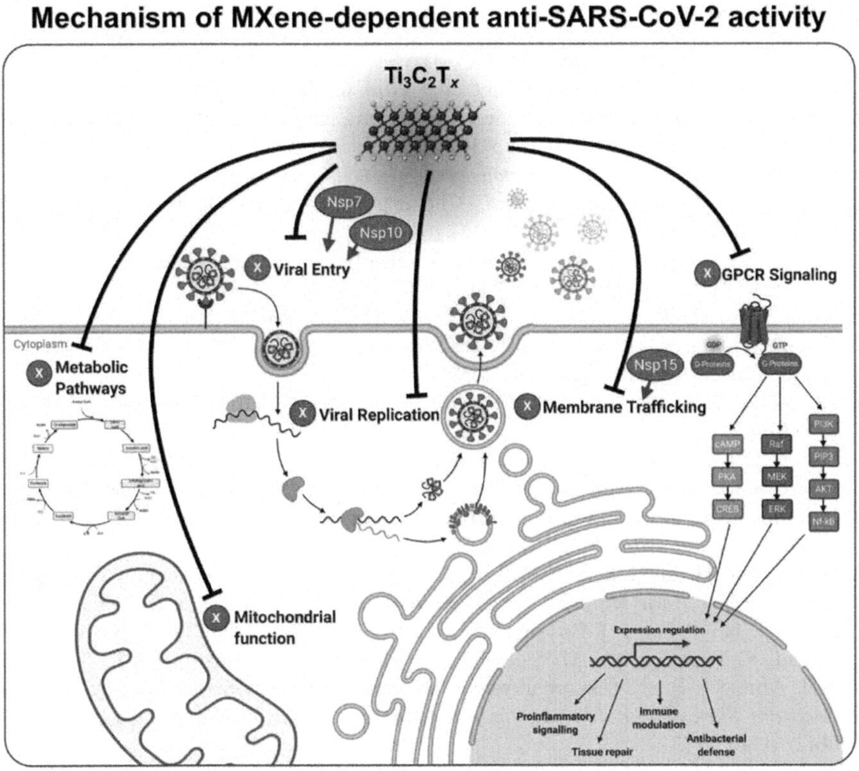

Mechanism of MXene-dependent anti-SARS-CoV-2 activity

FIGURE 4.4 The schematic showing the possible modes of antiviral action of MXene and related materials for the SARS-CoV-2 virus.

Source: Reproduced with permission from ref. [48].

MXene also depends upon the virus genotype and mutations. $Ti_3C_2T_x$ was the only MXene that could significantly inhibit infection in SARS-CoV-2/clade GR infected Vero E6 cells. $Ti_3C_2T_x$ demonstrated viral inhibitory effectiveness not just at the cell surface but also through several regulatory channels, including membrane transportation, mitochondrial function, metabolic functions, G protein–coupled receptor (GPCR) signaling, and multiplication of virus. GNG5, GRPEL1, and NUTF2 were host proteins that were important regulators of $Ti_3C_2T_x$-dependent antiviral effectiveness. These proteins of the host cells interact with the virus proteins (NSP7, NSP10, and NSP15). During the SARS-CoV-2 infection phase, NSP7 protein plays a significant role in GPCR signaling and membrane trafficking. NSP15 was engaged both in transportation of vehicles and nuclear transportation gear. NSP7 and NSP10 proteins can alter endomembrane to facilitate virus entrance and reproduction. Furthermore, immunological analysis of MXenes revealed the material's high biocompatibility and immune compliance, as well as its ability to block monocytes and reduce the production of proinflammatory cytokines, implying that these 2D structures have an anti-inflammatory effect. Figure 4.4 shows the proposed mechanism of antiviral action for $Ti_3C_2T_x$ MXenes.

REFERENCES

1. Howard, S.J., S. Hopwood, and S.C. Davies, *Antimicrobial Resistance: A Global Challenge.* Science Translational Medicine, 2014. **6**(236): p. 236ed10.
2. Sugden, R., R. Kelly, and S. Davies, *Combatting Antimicrobial Resistance Globally.* Nature Microbiology, 2016. **1**(10): p. 16187.
3. Sun, W. and F.G. Wu, *Two-Dimensional Materials for Antimicrobial Applications: Graphene Materials and Beyond.* Chemistry—An Asian Journal, 2018. **13**(22): p. 3378–3410.
4. Rasool, K., M. Helal, A. Ali, C.E. Ren, Y. Gogotsi, and K.A. Mahmoud, *Antibacterial Activity of Ti3C2T x MXene.* ACS Nano, 2016. **10**(3): p. 3674–3684.
5. Arabi Shamsabadi, A., M. Sharifian Gh., B. Anasori, and M. Soroush, *Antimicrobial Mode-of-Action of Colloidal Ti3C2T x MXene Nanosheets.* ACS Sustainable Chemistry & Engineering, 2018. **6**(12): p. 16586–16596.
6. Novoselov, K.S. and A. Geim, *The Rise of Graphene.* Nature Materials, 2007. **6**(3): p. 183–191.
7. Ferrari, A.C., et al., *Science and Technology Roadmap for Graphene, Related Two-Dimensional Crystals, and Hybrid Systems.* Nanoscale, 2015. **7**(11): p. 4598–4810.
8. Pasricha, R., S. Gupta, and A.K. Srivastava, *A Facile and Novel Synthesis of Ag-Graphene-based Nanocomposites.* Small, 2009. **5**(20): p. 2253–2259.
9. Weiss, N.O., H. Zhou, L. Liao, Y. Liu, S. Jiang, Y. Huang, and X. Duan, *Graphene: An Emerging Electronic Material.* Advanced Materials, 2012. **24**(43): p. 5782–5825.
10. Dwivedi, N., T. Patra, J.-B. Lee, R.J. Yeo, S. Srinivasan, T. Dutta, K. Sasikumar, C. Dhand, S. Tripathy, M.S.M. Saifullah, A. Danner, S.A.R. Hashmi, A.K. Srivastava, J.-H. Ahn, S.K.R.S. Sankaranarayanan, H. Yang, and C.S. Bhatia, *Slippery and Wear Resistant Surfaces Enabled by Interface Engineered Graphene.* Nano Letters, 2019. **20**(2): p. 905–917.
11. Mukherjee, M.D., C. Dhand, N. Dwivedi, B.P. Singh, G. Sumana, V.V. Agarwal, J.S. Tawale, and B.D. Malhotra, *Facile Synthesis of 2-Dimensional Transparent Graphene Flakes for Nucleic Acid Detection.* Sensors and Actuators B: Chemical, 2015. **210**: p. 281–289.
12. Devi, R., S. Gogoi, H.S. Dutta, M. Bordoloi, S.K. Sanghi, and R. Khan, *Au/NiFe₂O₄ Nanoparticle-Decorated Graphene Oxide Nanosheets for Electrochemical Immunosensing of Amyloid Beta Peptide.* Nanoscale Advances, 2020. **2**: p. 239–248.
13. Bharti, D.K., M.K. Gupta, R. Kumar, N. Sathish, and A.K. Srivastava, *Non-Centrosymmetric Zinc Silicate-Graphene based Transparent Flexible Piezoelectric Nanogenerator.* Nano Energy, 2020: p. 104821.
14. Tiwari, J.K., A. Mandal, N. Sathish, A.K. Agrawal, and A.K. Srivastava, *Investigation of Porosity, Microstructure and Mechanical Properties of Additively Manufactured Graphene Reinforced AlSi10Mg Composite.* Additive Manufacturing, 2020. **33**: p. 101095.
15. Karahan, H.E., C. Wiraja, C. Xu, J. Wei, Y. Wang, L. Wang, F. Liu, and Y. Chen, *Graphene Materials in Antimicrobial Nanomedicine: Current Status and Future Perspectives.* Advanced Healthcare Materials, 2018. **7**(13): p. 1701406.
16. Zou, X., L. Zhang, Z. Wang, and Y. Luo, *Mechanisms of the Antimicrobial Activities of Graphene Materials.* Journal of the American Chemical Society, 2016. **138**(7): p. 2064–2077.
17. Wu, F., S. Zhao, B. Yu, Y.-M. Chen, W. Wang, Z.-G. Song, Y. Hu, Z.-W. Tao, J.-H. Tian, Y.-Y. Pei, M.-L. Yuan, Y.-L. Zhang, F.-H. Dai, Y. Liu, Q.-M. Wang, J.-J. Zheng, L. Xu, E.C. Holmes, and Y.-Z. Zhang, *A New Coronavirus Associated with Human Respiratory Disease in China.* Nature, 2020. **579**(7798): p. 265–269.
18. Stadnytskyi, V., C.E. Bax, A. Bax, and P. Anfinrud, *The Airborne Lifetime of Small Speech Droplets and Their Potential Importance in SARS-CoV-2 Transmission.* Proceedings of the National Academy of Sciences, 2020. **117**(22): p. 11875–11877.

19. Van Doremalen, N., T. Bushmaker, D.H. Morris, M.G. Holbrook, A. Gamble, B.N. Williamson, A. Tamin, J.L. Harcourt, N.J. Thornburg, S.I. Gerber, J.O. Lloyd-Smith, E. de Wit, and V.J. Munster, *Aerosol and Surface Stability of SARS-CoV-2 as Compared with SARS-CoV-1*. New England Journal of Medicine, 2020. **382**(16): p. 1564–1567.

20. Chin, A., J.T.S. Chu, M.R.A. Perera, K.P.Y. Hui, H.-L. Yen, M.C.W. Chan, M. Peiris, and L.L.M. Poon, *Stability of SARS-CoV-2 in Different Environmental Conditions*. Lancet Microbe, 2015. **1**: p. e10.

21. Perreault, F., A.F. de Faria, S. Nejati, and M. Elimelech, *Antimicrobial Properties of Graphene Oxide Nanosheets: Why Size Matters*. ACS Nano, 2015. **9**(7): p. 7226–7236.

22. Hu, W., C. Peng, W. Luo, M. Lv, X. Li, D. Li, Q. Huang, and C. Fan, *Graphene-Based Antibacterial Paper*. ACS Nano, 2010. **4**(7): p. 4317–4323.

23. Li, C., X. Wang, F. Chen, C. Zhang, X. Zhi, K. Wang, and D. Cui, *The Antifungal Activity of Graphene Oxide—Silver Nanocomposites*. Biomaterials, 2013. **34**(15): p. 3882–3890.

24. Sametband, M., I. Kalt, A. Gedanken, and R. Sarid, *Herpes Simplex Virus Type-1 Attachment Inhibition by Functionalized Graphene Oxide*. ACS Applied Materials & Interfaces, 2014. **6**(2): p. 1228–1235.

25. Ye, S., K. Shao, Z. Li, N. Guo, Y. Zuo, Q. Li, Z. Lu, L. Chen, Q. He, and H. Han, *Antiviral Activity of Graphene Oxide: How Sharp Edged Structure and Charge Matter*. ACS Applied Materials & Interfaces, 2015. **7**(38): p. 21571–21579.

26. Ziem, B., H. Thien, K. Achazi, C. Yue, D. Stern, K. Silberreis, M. Fardin Gholami, F. Beckert, D. Gröger, R. Mülhaupt, J.P. Rabe, A. Nitsche, and R. Haag, *Highly Efficient Multivalent 2D Nanosystems for Inhibition of Orthopoxvirus Particles*. Advanced Healthcare Materials, 2016. **5**(22): p. 2922–2930.

27. Ziem, B. and W. Azab, *Size-Dependent Inhibition of Herpesvirus Cellular Entry by Polyvalent Nanoarchitectures*, 2017. **9**(11): p. 3774–3783.

28. Chen, Y.N., Y.-H. Hsueh, C.-T. Hsieh, D.-Y. Tzou, and P.-L. Chang, *Antiviral Activity of Graphene-Silver Nanocomposites against Non-Enveloped and Enveloped Viruses*. International Journal of Environmental Research and Public Health, 2016. **13**(4): p. 430.

29. Frost, R., G.E. Jönsson, D. Chakarov, S. Svedhem, and B. Kasemo, *Graphene Oxide and Lipid Membranes: Interactions and Nanocomposite Structures*. Nano Letters, 2012. **12**(7): p. 3356–3362.

30. Raval, B. and A.K. Srivastav, *Synthesis of Exfoliated Multilayer Graphene and Its Putative Interactions with SARS-CoV-2 Virus Investigated through Computational Studies*. Journal of Biomolecular Structure and Dynamics, 2022. **40**(2): p. 712–721.

31. Łoczechin, A., K. Séron, A. Barras, E. Giovanelli, S. Belouzard, Y.-T. Chen, N. Metzler-Nolte, R. Boukherroub, J. Dubuisson, and S. Szunerits, *Functional Carbon Quantum Dots as Medical Countermeasures to Human Coronavirus*. ACS Applied Materials & Interfaces, 2019. **11**(46): p. 42964–42974.

32. Barras, A., Q. Pagneux, F. Sane, Q. Wang, R. Boukherroub, D. Hober, and S. Szunerits, *High Efficiency of Functional Carbon Nanodots as Entry Inhibitors of Herpes Simplex Virus Type 1*. ACS Applied Materials & Interfaces, 2016. **8**(14): p. 9004–9013.

33. Seifi, T. and A. Reza Kamali, *Antiviral Performance of Graphene-based Materials with Emphasis on COVID-19: A Review*. Medicine in Drug Discovery, 2021. **11**: p. 100099.

34. Yang, X.X., C.M. Li, Y. Fang Li, J. Wang, and C. Zhi Huang, *Synergistic Antiviral Effect of CurCumin Functionalized Graphene Oxide against Respiratory Syncytial Virus Infection*. Nanoscale, 2017. **9**(41): p. 16086–16092.

35. Du, T., J. Lu, L. Liu, N. Dong, L. Fang, S. Xiao, and H. Han, *Antiviral Activity of Graphene Oxide—Silver Nanocomposites by Preventing Viral Entry and Activation of the Antiviral Innate Immune Response*. ACS Applied Bio Materials, 2018. **1**(5): p. 1286–1293.

36. Elechiguerra, J.L., J.L. Burt, J.R. Morones, A. Camacho-Bragado, X. Gao, H.H Lara, and M.J. Yacaman, *Interaction of Silver Nanoparticles with HIV-1*. Journal of Nanobiotechnology, 2005. **3**: p. 6.

37. Deokar, A.R., A.P. Nagvenkar, I. Kalt, L. Shani, Y. Yeshurun, A. Gedanken, and R. Sarid, *Graphene-Based "Hot Plate" for the Capture and Destruction of the Herpes Simplex Virus Type 1.* Bioconjugate Chemistry, 2017. **28**(4): p. 1115–1122.

38. Song, Z., X. Wang, G. Zhu, Q. Nian, H. Zhou, D. Yang, C. Qin, and R. Tang, *Virus Capture and Destruction by Label-Free Graphene Oxide for Detection and Disinfection Applications.* Small, 2015. **11**(9–10): p. 1171–1176.

39. Ting, D., N. Dong, L. Fang, J. Lu, J. Bi, S. Xiao, and H. Han, *Multisite Inhibitors for Enteric Coronavirus: Antiviral Cationic Carbon Dots Based on Curcumin.* ACS Applied Nano Materials, 2018. **1**(10): p. 5451–5459.

40. Das Jana, I., P. Kumbhakar, SA. Banerjee, and C. Chowde Gowda, *Copper Nanoparticle— Graphene Composite-Based Transparent Surface Coating with Antiviral Activity against Influenza Virus.* ACS Applied Nano Materials, 2021. **4**(1): p. 352–362.

41. Gogotsi, Y. and B. Anasori, *The Rise of MXenes.* ACS Nano, 2019. **13**(8): p. 8491–8494.

42. Huang, K., Z. Li, J. Lin, G. Han and P. Huang, *Two-Dimensional Transition Metal Carbides and Nitrides (MXenes) for Biomedical Applications.* Chemical Society Reviews, 2018. **47**(14): p. 5109–5124.

43. Dai, C., Y. Chen, X. Jing, L. Xiang, D. Yang, H. Lin, Z. Liu X. Han, and R. Wu, *Two-Dimensional Tantalum Carbide (MXenes) Composite Nanosheets for Multiple Imaging-Guided Photothermal Tumor Ablation.* ACS Nano, 2017. **11**(12): p. 12696–12712.

44. Murphy, B.B., P.J. Mulcahey, N. Driscoll, A.G. Richardson, G.T. Robbins, N.V. Apollo, K. Maleski, T.H. Lucas, Y. Gogotsi, T. Dillingham, and F. Vitale, *A Gel-free Ti(3)C(2)T(x)-based Electrode Array for High-Density, High-Resolution Surface Electromyography.* Advanced Materials Technologies, 2020. **5**(8).

45. Rasool, K., M. Helal, A. Ali, C.E. Ren, Y. Gogotsi, and K.A. Mahmoud, *Antibacterial Activity of Ti3C2Tx MXene.* ACS Nano, 2016. **10**(3): p. 3674–3684.

46. Rasool, K., K.A. Mahmoud, D.J. Johnson, M.Helal, G.R. Berdiyorov, and Y. Gogotsi, *Efficient Antibacterial Membrane based on Two-Dimensional Ti(3)C(2)T(x) (MXene) Nanosheets.* Scientific Reports, 2017. **7**(1): p. 1598–1598.

47. Pandey, R.P., K. Rasool, V.E. Madhavan, B. Aïssa, Y. Gogotsi, and K.A. Mahmoud, *Ultrahigh-Flux and Fouling-Resistant Membranes Based on Layered Silver/MXene (Ti3C2Tx) Nanosheets.* Journal of Materials Chemistry A, 2018. **6**(8): p. 3522–3533.

48. Unal, M.A., F. Bayrakdar, L. Fusco, O. Besbinar, C.E. Shuck, S. Yalcin, M. Turktas Erken, A. Ozkul, C. Gurcan, O. Panatli, G.Y. Summak, C. Gokce, M. Orecchioni, A. Gazzi, F. Vitale, J. Somers, E. Demir, S.S. Yildiz, H. Nazir, J.-C. Grivel, D. Bedognetti, A. Crisanti, K.C. Akcali, Y. Gogotsi, L.G. Delogu, and A. Yilmazer, *2D MXenes with Antiviral and Immunomodulatory Properties: A Pilot Study against SARS-CoV-2.* Nano Today, 2021. **38**: p. 101136.

5 Advanced Biosensors Based on 2D Materials for COVID-19

Manoj Kumar Gupta

The COVID-19 pandemic, which began as a respiratory infection, has created severe damage to society, causing panic around the world [1, 2]. As compared to other deadly viral outbreaks, the severity of the severe acute respiratory syndrome corona-virus 2 (SARS-CoV-2) was initially deeply overlooked; however, it quickly became a global public health concern, and about a total of 500 million cases have been reported in April 2022, while the death toll currently stands to around 6 million. To prevent the spread of the COVID-19 virus, an efficient technique is required [3, 4]. In this regard, detection of COVID-19 and other similar viruses through a simple and rapid strategy has emerged as an effective tool, and various methods have been recently developed [5, 6]. The accurate detection of COVID-19 is one of the important steps for the quick isolation and treatment of COVID-19-infected patients [7, 8]. Particularly, biosensors are an analytical device that converts biological response into quantifiable and processable signals [9]. The measurable signal can be in the form of electrical, electrochemical, optical, or mechanical, and the type of the signal determines the type of the biosensor [10–12]. These biosensors have been proven as promising tools for the early-stage diagnosis of several infections, including from the COVID-19 virus [9, 13]. To develop the biosensors, recently, various types of nanostructures, including graphene, graphene oxide, graphene quantum dots, MoS_2, MXene, carbon nanotubes, piezoelectric nanostructures (including ZnO, PZT, and $BaTiO_3$), have been widely used for the early-stage diagnosis of infections, including viral infections and SARS [14–19]. Due to the advantage of tailoring the specific interaction of compounds by immobilizing biological recognition materials on the sensor substrate, the performance and sensitivity of the biosensors have improved, as it has specific binding affinity to the desired molecule. Particularly, graphene has received tremendous attraction for designing the highly sensitive biosensors due to its outstanding electrical, mechanical, and chemical properties, high surface area, and thermal and electrical conductivity [20, 21]. Graphene is a one-atom-thick two-dimensional (2D) sheet of carbon atoms with sp2 hybridization arranged in a honeycomb lattice structure, and the charge carriers inside the graphene act as the "massless" particles [22–24]. Due to its high surface area, graphene has been one of the most interesting materials to develop electrodes and active materials for the sensor. It has been used in different forms, including nanoparticle and oxide forms [25–27]. Graphene has been used actively used in flexible and stretchable sensors,

DOI: 10.1201/9781003316381-5

electrochemical sensors, and flexible biosensors. In addition, transition metal dichalcogenides (TMDs) have also emerged as analogues of the graphene, with outstanding physiochemical and electronic properties for advanced point-of-care diagnosis of COVID-19 virus [28, 29]. Numerous types of device structures are being explored toward the fabrication and detection of electrical sensing using 2D materials, including chemical diodes, capacitors, and field-effect transistors (FETs). Particularly, field effect transistor (FET) and metal-oxide-semiconductor FET (MOSFET) devices have also gained a lot of attraction due to their compatibility with point-of-care applications and have received interest as they can easily be integrable with existing semiconductor manufacturing technologies [30–32]. Field-effect transistors (FETs) based on reduced graphene, graphene oxide, or graphene amine have been widely used for the detection of DNA hybridization, negatively charged bacteria, and immunoglobulin G and SaRS-COV-2 [33–35]. The gate of the FET devices has been used to control carrier mobilities to achieve a higher sensitivity toward various virus [36].

5.1   GRAPHENE-BASED FET BIOSENSORS

Yasuhide Ohno and his team have developed label-free biosensors based on aptamer-modified graphene field-effect transistors. In this work, they have reported a 2D graphene–based FET biological sensing applications, where the performance of the FET depends on two parameters: first one is the Debye length, and the other is the functionalization of the receptor [23]. Debye length is the typical distance required for screening the surplus charge by mobile carriers, which usually depends on ionic strength and temperature. The mobile charges in a transistor's channel are not affected by charged molecules located more than a Debye length away. They have reported that a receptor–ligand reaction must occur within the Debye length. The benefit of using aptamers in their experiments is that they are smaller than the Debye length. The functionalization of the receptor without introducing defects on the graphene surface for enabling specific detection and for minimizing the nonspecific binding is also one of the key factors to enhance performance. Their research team has fabricated the graphene FET with functionalization of IgE aptamers. To enable IgE sensing, the IgE aptamers were immobilized on single-layer graphene. The atomic-force microscopy (AFM) and electrical measurements were carried out to confirm the sensing, where the target protein was an antibody immunoglobulin E (IgE) [37, 38]. The height profile of the surface was evaluated by AFM topographical image. The height of the IgE aptamers was about ~3 nm (**Figure 5.1a, b**). Based on the heigh profile, the authors suggested that protein–aptamer reactions are expected to occur inside the electrical double layer. The aptamer-modified graphene-FET electrically detected only for the IgE protein [39]. The AFM result shows the immobilization of IgE aptamers on the graphene surface, which were carried out before and after the functionalization (Figure 5.1). Before functionalization with IgE aptamers, a single layer with 0.3–0.5 nm thick graphene channel was observed. However, after functionalization, the height of the channel is increased to ~3 nm. The results confirmed that IgE aptamers were immobilized only on the graphene surface [23]. The electrical characteristics were also investigated to observe the effects of functionalizing on graphene FET (**Figure 5.1c**), and it was found that the IgE aptamers were

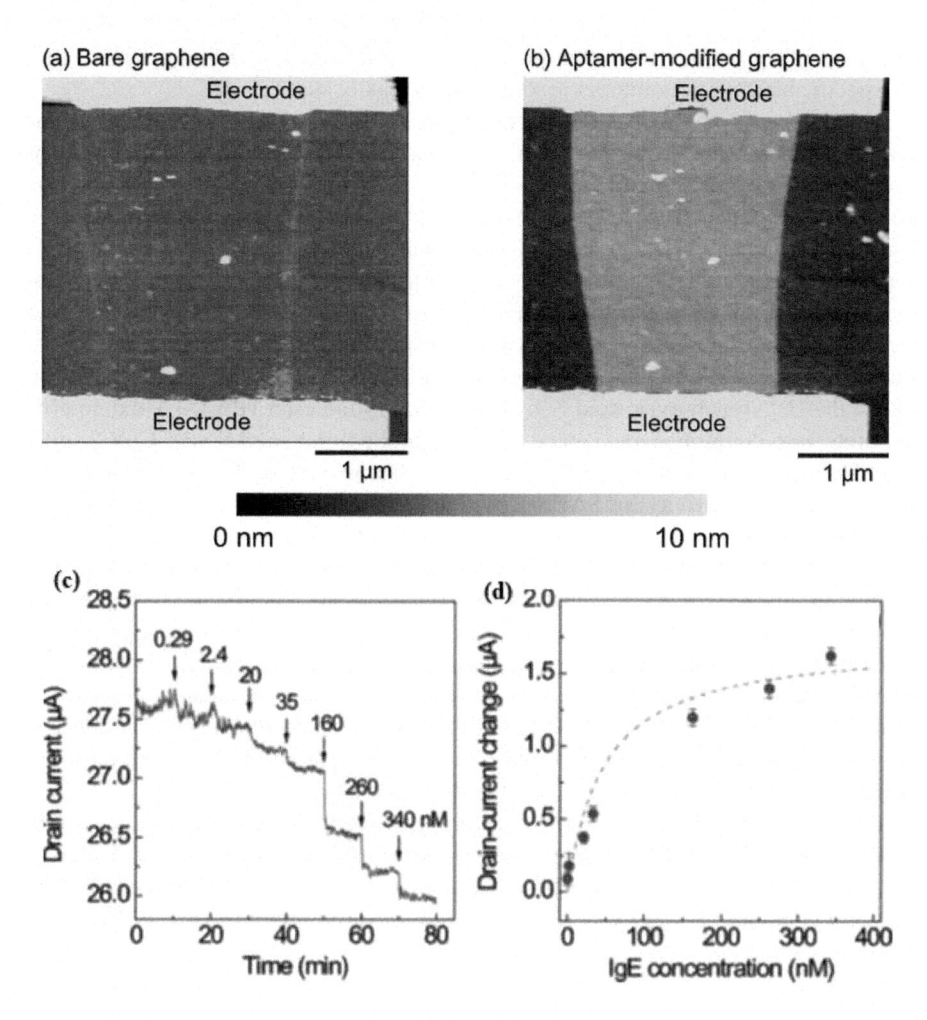

FIGURE 5.1 (a) AFM image of the bare graphene channel of the FET and (b) aptamer-modified graphene channel of the FET. (c) Time course of the drain current (I_d) aptamer-modified graphene FET. (d) Change in the drain current with the IgE concentration.

successfully immobilized in the graphene channel and that the graphene-based FET can electrically detect the existence of oligonucleotides on its surface.

The real-time electrical performance of the graphene FET with the target IgE protein at various concentrations of 0.29, 2.4, 20, 35, 160, 260, and 340 nm into an aptamer-modified graphene FET was monitored. The drain current was decreased stepwise after injection of the target IgE at each concentration (**Figure 5.1c**). The net change in drain current with the function of IgE concentration was also recorded, and results indicate an increasing trend with increasing IgE concentration; however, it gradually saturated above concentrations of 160 nM (**Figure 5.1d**). The results demonstrated that the aptamer-modified 2D graphene FET electrically can easily detect the IgE protein, while other proteins are not detected, and findings show that nonspecific binding of nontarget proteins can be easily suppressed.

A novel graphene-based field-effect transistor (FET) based biosensing device for detecting severe acute respiratory syndrome coronavirus 2 (SARS-CoV-2) has been developed by Giwan Seo and his research team [33]. As COVID-19 is an infectious disease associated with severe respiratory distress, to detect the COVID-19 virus, crystalline graphene layer–based FET functionalized with SARS-CoV-2 spike antibody was constructed. The biosensor was fabricated using the coating of graphene sheets on the FET with a specific antibody against SARS-CoV-2 spike protein. The performance of the sensor was determined using antigen protein, cultured virus, and nasopharyngeal swab specimens collected from COVID-19 patients. They have shown that 1-pyrenebutyric acid N-hydroxysuccinimide ester (PBASE) and an efficient interface coupling agent can be used as a probe linker, and SARS-CoV-2 spike antibody was immobilized onto the graphene-based device. The developed FET device exhibits sensitivity for SARS-CoV-2 antigen protein with a limit of detection (LOD) of 1 fg/mL.

Real-time detection of SARS-CoV-2 antigen protein and of SARS-CoV-2 virus through graphene-based FET sensor was also performed. The dynamic response of the sensor to spike protein was recorded to examine the performance of graphene-based FET sensor (**Figure 5.2**). Graphene surface is chemically functionalized using PBASE, which was confirmed through the Raman spectra and X-ray photoelectron spectroscopy (XPS). The graphene-based FET sensor showed a very lower concentration even for the 1 fg/mL of SARS-CoV-2 spike protein in phosphate-buffered saline (PBS), which is better than LOD of the ELISA platform.

The graphene FET device fabricated with pristine graphene in the absence of the SARS-CoV-2 spike protein conjugation does not exhibited any significant change in the signal after the introduction of various sample concentrations (**Figure 5.3**). The response behavior of the device was measured for other proteins as well. However,

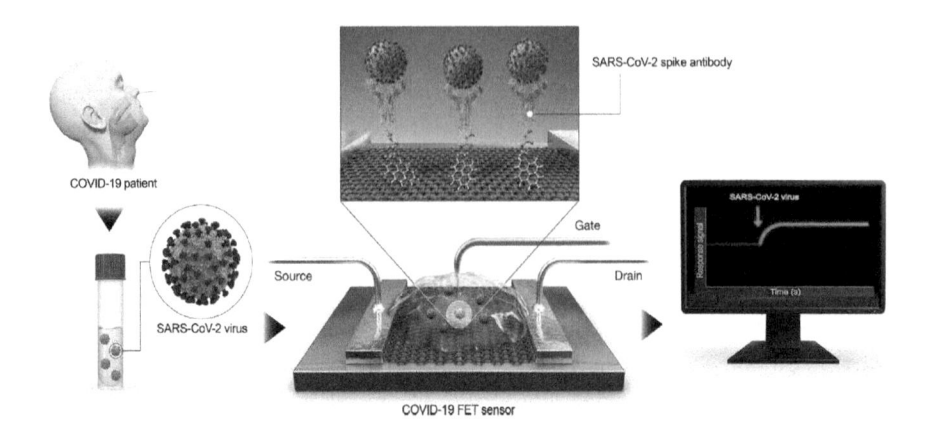

FIGURE 5.2   The process diagram of the graphene-based FET sensor for COVID-19 detection. SARS-CoV-2 is attached via 1-pyrenebutyric acid N-hydroxysuccinimide ester on the graphene layer.

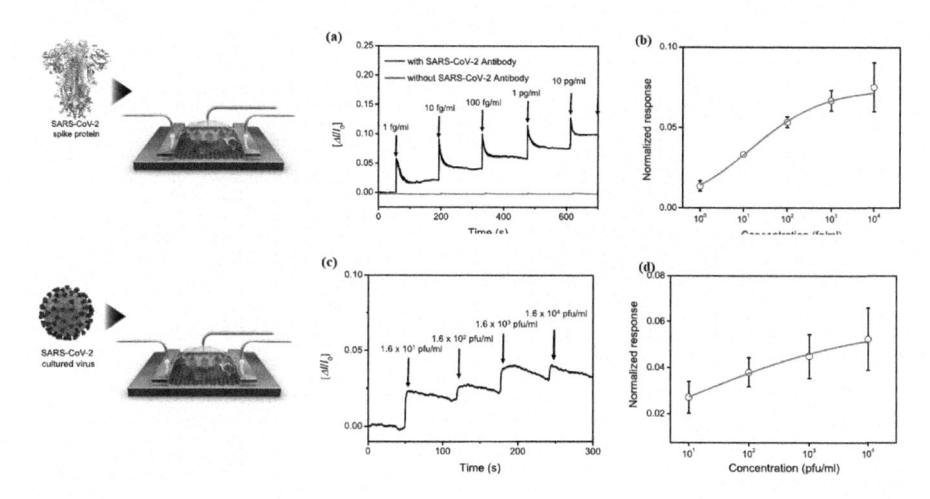

FIGURE 5.3 (a) Schematic image for the detection of SARS-CoV-2 antigen protein through the COVID-19 FET sensor for SARS-CoV-2 spike protein. (b) Real-time response for SARS-CoV-2 antigen protein. (c) Detection of cultured SARS-CoV-2 virus. (d) Dose-dependent response curve for SARS-CoV-2 virus.

the graphene-based COVID-19 FET sensor has not shown any response to other proteins, such as MERS-CoV, which confirmed that the COVID-19 graphene FET sensor has high sensitivity and specificity for the SARS-CoV-2 spike antigen protein. Moreover, for real-time detection of the COVID-19 virus performed, SARS-CoV-2 was propagated in cultured cells; the COVID-19 FET sensor device showed significant response even for low concentrations of 1.6×10^1 pfu/mL. Further, a graphene sensor was able to detect SARS-CoV-2 virus in clinical samples. Their results showed that the COVID-19 graphene FET sensor is an excellent candidate for COVID-19 diagnosis. The functionalized graphene-based FET sensor platform enables a simple, rapid, and highly responsive detection system for the SARS-CoV-2 virus in clinical samples.

Epitaxial graphene nanosheets developed on silicon carbide have several advantages over CVD-grown graphene for application in graphene-based electronics. A bilayer quasi-freestanding epitaxial graphene (QFS EG) has also shown great potential for the detection of the SARS-CoV-2 virus. High-quality graphene is synthesized via Si sublimation and hydrogen intercalation on a 4 in diameter semi-insulating (0001) ~0.1° off-axis 6H-SiC using a horizontal hot-wall reactor. Bilayer graphene was used as an active layer for the rapid detection of SARS-CoV-2 in infected patients, where exhaled breath aerosol samples and mid-turbinate swab samples are used to measure the performance of the epitaxial 2D graphene–based biosensor device for low concentration. The SARS-CoV-2 S1 spike protein antibodies are immobilized on bilayer quasi-freestanding epitaxial graphene (**Figure 5.4**). The SARS-CoV-2 samples are detected via this device, and a very low concentration of 60 copies/mL of the SARS-CoV-2 virus was detected in seconds by electrical transduction of the SARS-CoV-2 S1 spike protein antigen. To fabricate graphene-based COVID-19

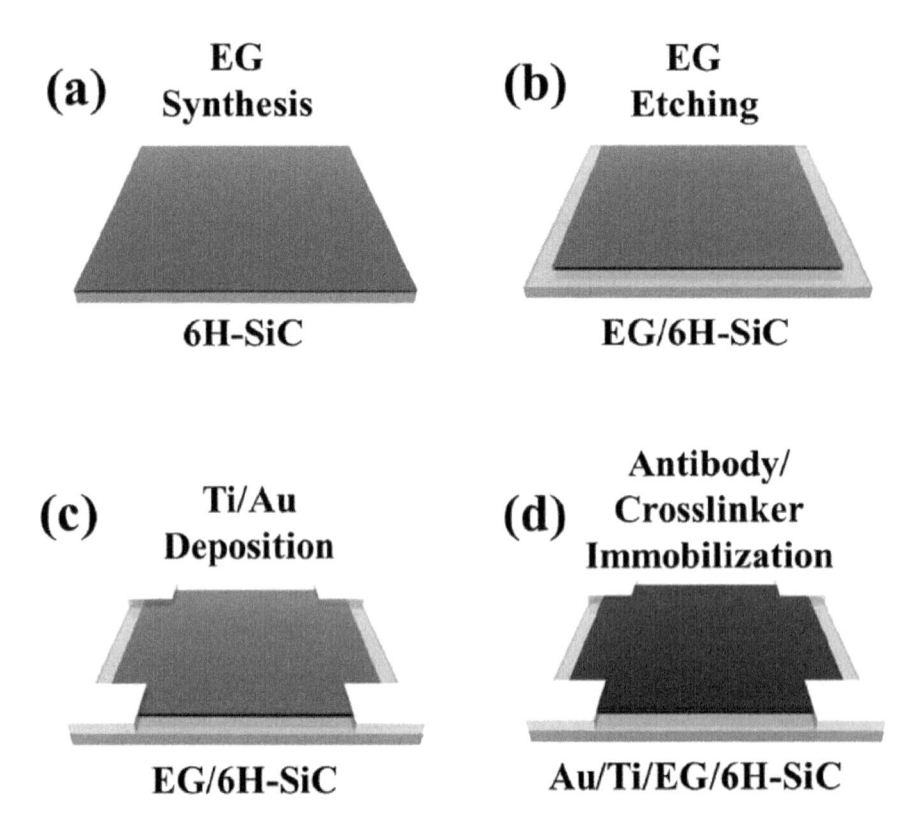

FIGURE 5.4 (a) Schematic of the synthesis process of quasi-freestanding bilayer epitaxial graphene on silicon carbide and (b) fabrication process of graphene-based COVID-19 sensor fabrication process. (c) Deposition of Ti/Au metal stack as electrodes using an e-beam evaporator. (d) Immobilization of crosslinker and SARS-CoV-2 S1 spike protein on EG.

biosensor, separately, optimized immobilization conditions of SARS-CoV-2 spike S1 protein antibody with crosslinker (poly-L-lysine) and hydrogen-intercalated QFS EG on silicon carbide (SiC) were used. The poly-L-lysine crosslinker helps create strong adhesion with epitaxial graphene through the hydrophobic interaction between its butyl chains and graphene surface for further functionalization with other bioactive materials. The conducting and thin layer of the Ti/Au contact electrodes is used for the graphene sensors. The AFM images of SARS-CoV-2 S1 spike protein antibody/crosslinker prepared on EG/SiC are measured under various condition. The AFM images show that immobilized antibody/crosslinker is dense and uniform on the graphene layer, with RMS value of 1.9 nm, while roughness of the crosslinker and EG-coated graphene sheets were 0.8 nm and 0.7 nm, respectively.

The performance of the epitaxial graphene sensor on the Si substrate under different SARS-CoV-2 S1 spike protein antigen concentrations, from 1 ag/mL to 1 μg/mL, is recorded. The spike protein solutions in ELISA assay diluent (1×) are prepared by serial dilution by dropwise incorporating 1 μL onto the graphene sensor, and the measurements were performed (Figure 5.5a,d). The graphene-based sensor exhibits a response as low as 1 ag/mL of SARS-CoV-2 S1 spike protein antigen in ELISA assay diluent, which was also one of the best LOD values reported on the biosensors. To confirm the performance of the COVID-19 sensor, the authors have tested the graphene-based device with 1 ag/mL of SARS-CoV-2 S1 spike protein antigen, and it was clearly found that the graphene sensor distinguished between a blank and 1 ag/mL protein and yielded an average response of 31% (**Figure 5.5**). The results indicate that an epitaxial 2D graphene–based COVID-19 sensor is highly sensitive and selective for the SARS-CoV-2 S1 spike protein antigen. The developed COVID-19 graphene sensor was demonstrated for the practical diagnosis application. The SARS-CoV-2 virus from infected patient samples, such as midturbinate swabs, saliva, exhaled breath aerosol samples, and common human coronaviruses are used to detect from the sensors. The responses of the COVID-19 sensor subjected to negative samples (tested negative by RT-PCR) and different concentrations of positive samples (swab), such as 60, 125, and 250 copies/mL of SARS-CoV-2 virus, were observed (**Figure 5.5e-f**). The epitaxial graphene-based biosensor can distinguish between negative and positive samples. The electric response increases monotonically with an increase in the concentration. The portable sensor demonstrated an exceptional SNR of 67.57 dB and a fast response time of 0.6 s.

DNA, a molecule which encodes genetic instructions, is known as the blueprint of life. The detection of the DNA has gained significant attraction because of its promising applications in various areas, including treatment and disease diagnosis, environmental monitoring, and detection of the COVID-19 virus. Normally, DNA detection technology is based on optical or electrochemical transductions, where fluorescent or electrochemical tags are essentially required. The label-free electrical detection has received huge attention nowadays due to its cost-effective and sequence-elective DNA sensors. Recently, a label-free detection of DNA hybridization using transistors based on CVD-grown graphene was developed by Tzu-Yin Chen *et al.* (**Figure 5.6a–d**) [24]. They have reported that due to the coexistence of metallic and semiconducting natures of carbon nanotubes, their application in biosensors is greatly restricted, and thus, performance of the DNA sensors varies significantly. To avoid this issue, 2D graphene nanosheet was used as metallic sheet, and graphene-based field-effect transistors based on CVD-grown large-area monolayer graphene on copper foils synthesized by chemical vapor deposition are utilized for label-free electrical detection of DNA hybridization. The liquid-gated FET device is constructed by making the source and drain electrodes on graphene using conductive silver paste. Researchers have measured the transfer curves (drain current I_d versus gate voltage V_g) of the liquid-gated FETs, whereas transferred and annealed graphene films are recorded in phosphate buffer edsaline (PBS) solutions. The charge neutrality point (VCNP: the applied gate voltage corresponding to the minimum conductance) from as-transferred and annealed graphene were 0.21 and 0.52 V, respectively. The influence of PBS concentration on the sensing performance of target DNA hybridized with probe DNA-decorated graphene FETs was examined. The 12-mer probe and target DNA strands were dissolved in 10×, 1×, and 0.× PBS solutions, respectively. The CVD graphene–based FET sensors have demonstrated DNA detection sensitivity of concentrations as low as 1 pM (10–12 M) (Figure 5.6

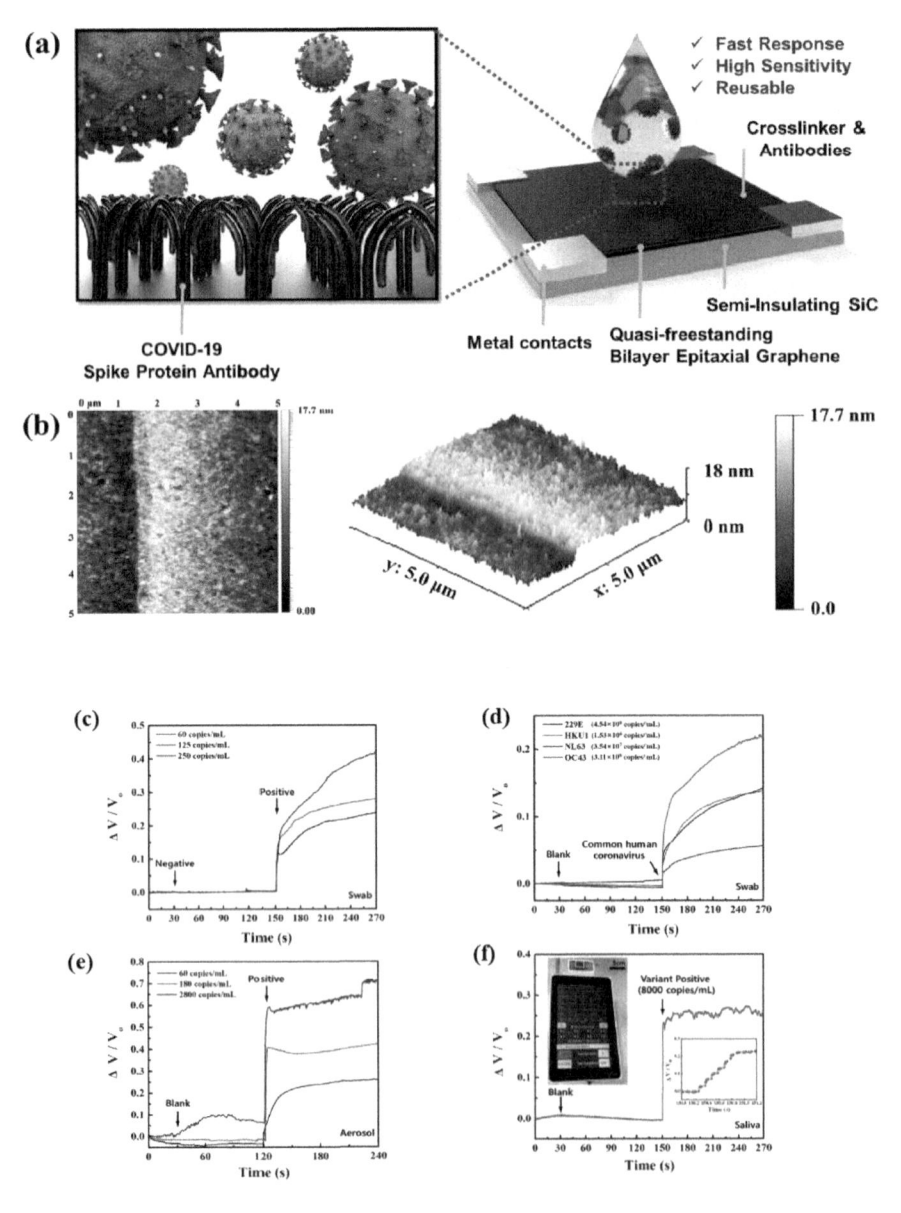

FIGURE 5.5 (a) Schematic diagram of the epitaxial graphene-based biosensor design. (b) AFM images of the surface on the antibody/crosslinker on graphene sheet layer. (c–d) Real-time testing and demonstration of a practical diagnosis application of the proposed COVID-19 sensor and response of common human coronaviruses using swabs on the COVID-19 sensor.

b, c). Moreover, the effect of the graphene surface was also analyzed, and it was also found that the VCNP shift is larger for Au-transferred graphene FET (Figure 5.6d), whereas performance of the graphene with PMMA coating was worse due to poor interaction between graphene and DNA due to PMMA coating.

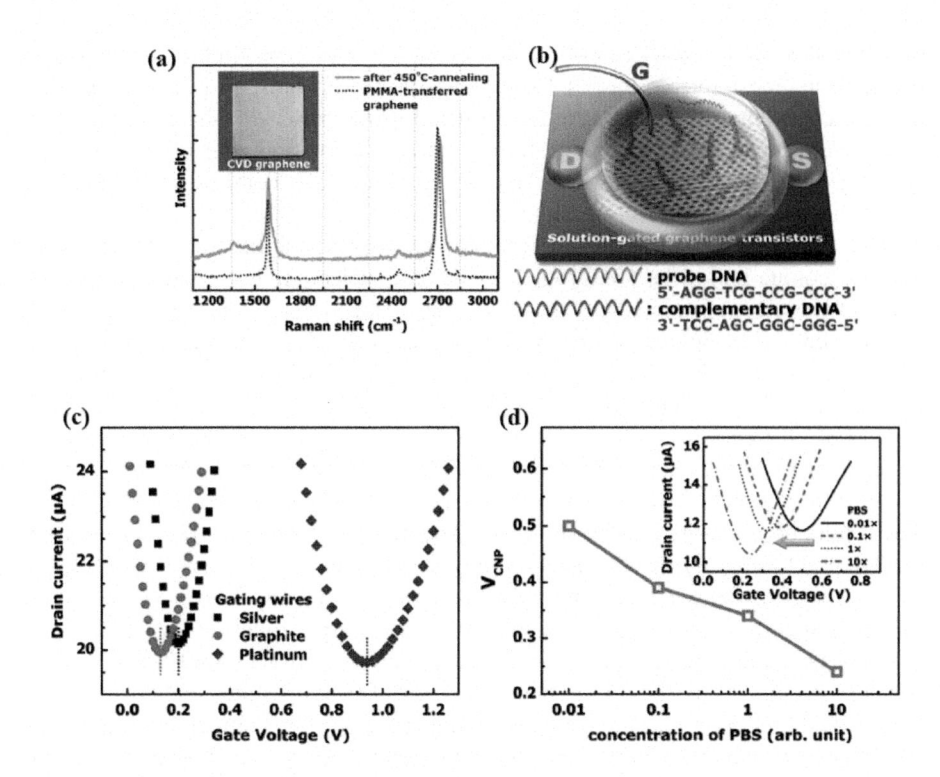

FIGURE 5.6 (a) Raman spectra of pure, annealed, and PMMA-based graphene nanosheets. (b) Schematic of solution-gated transistor. (c) Response of drain current for various electrode. (d) Variation of voltage with the PBS concentration

5.2 2D TRANSITION METAL DICHALCOGENIDES (TMD) BASED FET BIOSENSORS

Among various 2D materials, transition metal dichalcogenides (TMD), such as MoS_2, have received huge attention for the fabrication of biosensors because of their stable and distinct semiconducting behavior, piezoelectric properties, fluorescent behaviors, abundant reactive sites for redox reactions, high room temperature electron mobility, and its unusual properties, including high surface-to-volume ratio [18]. Various methods have been used to synthesize the monolayer, bilayer, and few-layer MoS_2 nanosheets, such as chemical vapor deposition, physical vapor deposition, and epitaxial, solgel, and hydrothermal routes [24]. Deblina Sarkar *et al.* have recently developed the MoS_2 field-effect transistor for next-generation label-free biosensors [18]. FET biosensors generally have two electrodes, namely, source and drain, which are connected to an MoS_2 semiconductor material (channel). The flow of current through the channel between the source and drain is electrostatically modulated by a third electrode, called the gate. In FET, the source and drain contacts are passivated with a dielectric layer to protect them from the electrolyte. In case of the not passivated, the source/drain metal electrodes come in contact with the electrolyte, and the biomolecules can be easily adsorbed directly on the electrodes, and thus it degrades the performance of the biosensors. In an MoS_2-based FET biosensor device, a fluidic channel containing the electrolyte is fabricated using an acrylic sheet, and an Ag/AgCl reference electrode as the electrolyte gate is used to apply bias to the electrolyte (Figure 5.7a–f). The MoS_2-based biosensor successfully demonstrated detection of pH changes of the electrolytic solution.

The mechanism of pH sensing from MoS_2 sensors was based on the protonation/deprotonation of the OH groups on the gate dielectric. The change in the pH value of the electrolyte changes the dielectric surface charge, and their responses are measured. The device performance in terms of the drain current as a function of the electrolyte gate voltage for various pH values of the electrolyte is measured (Figure 5.7b–d). The authors showed that the device exhibits a valuable increase in current at a certain applied bias, and a decrease in the pH value is obtained, which confirmed the successful demonstration of the MoS_2 pH sensor [18]. A biosensor which detects down to a single molecule is highly desirable (Figure 5.7e-f). The results of this work clearly show that detection of low-concentration detection (LCD) of 2D MoS_2 highly depends on site type of biomolecule, density on the effective layer, sensor area, analyte mass transport in the solution. A rigorous quantum mechanical simulation based on a nonequilibrium Green's function formalism as a function of applied biases for single-molecule detection from MoS2-based device was also developed by the researchers. The local density of states and current was obtained (Figure 5.7e–f). The results show that even at 5 nm channel length, the 2D MoS_2 has maintained excellent gate control over the channel, leading to near-ideal subthreshold swing (SS), which is essential to achieve high sensitivity. The MoS2 biosensor has various advantages over other materials, such as higher sensitivity for detection of single quanta of a biomolecular element, which facilitates low-power and high-density biosensor device architectures. Particularly, it was reported that MoS_2-based pH sensor exhibits ultrahigh sensitivity (713 for a pH change of 1 unit) as well as wide operation range (pH of 3 to 9). 2D MoS_2 devices have also shown potential for specific detection of protein with extremely high sensitivity of 196 [40].

The development of a rapid, highly sensitive, and specific assay for the rapid detection of antibodies to SARS-CoV-2 infection within seconds is highly needed for effective treatment. Detection through electrochemical methods and readout of biological molecule via a smartphone within early days of infection is very useful. In the electrochemical transduction method, the formation of antibody–antigen complex is detected; however, the sensitivity, specificity, and speed of detection will vary on the electrochemical cell and their geometry. The surface chemistry of the electrodes, antigen, and assay procedures also played an important role in the effective and efficient fabrication of the biosensors. Recently, Ali *et al.* have reported the 3D-printed test chip (3DcC) platform for detection of the COVID-19 virus. 3D printing techniques involving advanced materials and nanostructures have emerged as effective tools for development of the SARS-COV-2 biosensors due to the possibility of intricate geometries, custom microstructures, and material combinations [41]. 3D printing offers numerous exciting advantages over conventional technologies due to its simple two-step fabrication process controlled by computer-aided design (CAD) programs, customizability, and

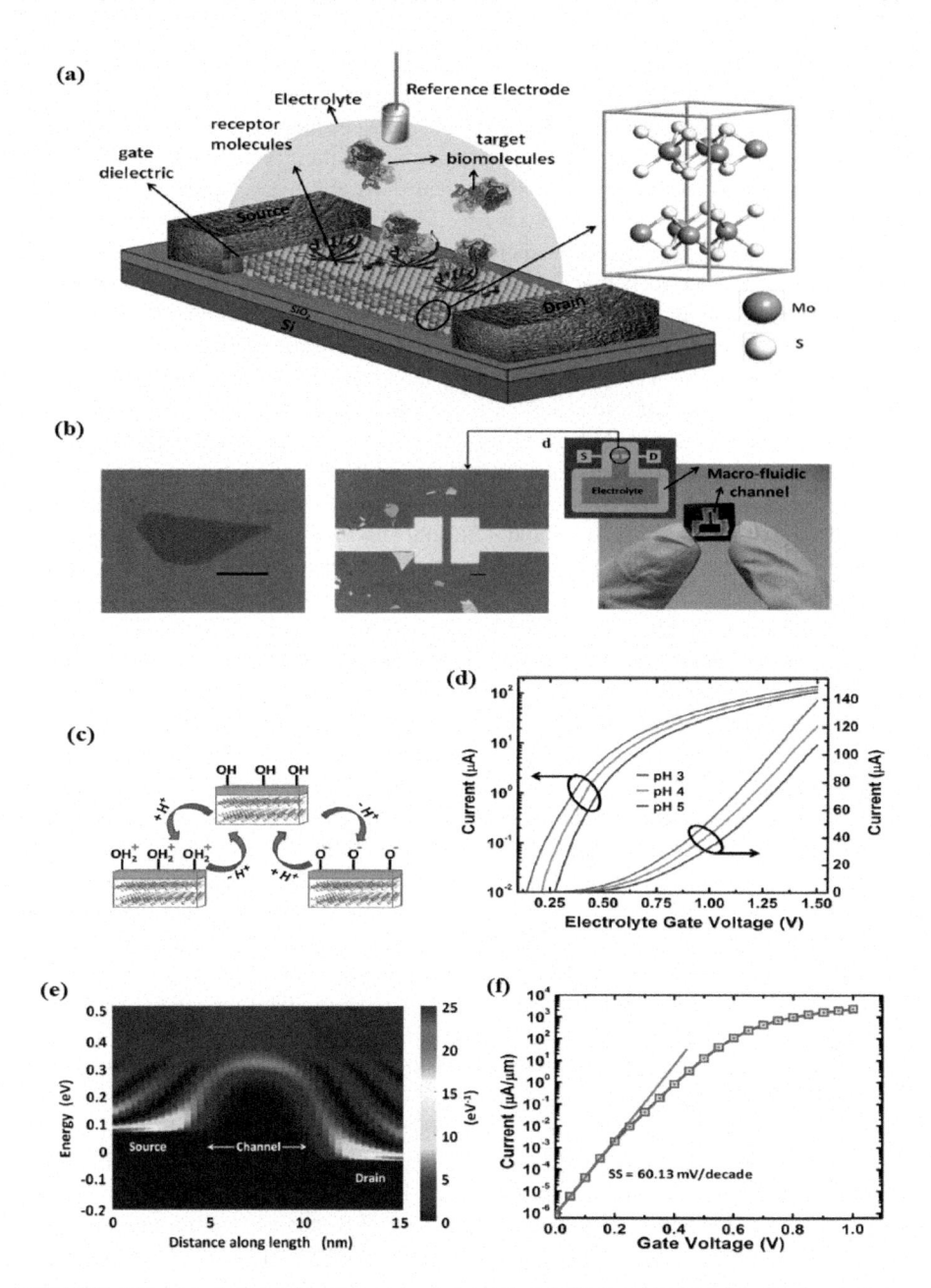

FIGURE 5.7 (a) Schematic diagram of the 2D MoS2–based FET biosensor device, which also shows an optical image of an MoS2 flake on SiO2/Si substrate with Ti/Au electrode used for pH sensing. (b) Protonated OH group on the dielectric surface to OH$_2$. (c) Drain current as a function of electrolyte gate voltage for three different pH values of the solution. (d) Simulation results of MoS$_2$-based FET sensors for local density of states and current.

prototypability. Specially, aerosol jet (AJ) 3D printing is a popular method which uses a stream of aerosolized droplets to deposit an array of nanomaterials with resolution of 10 μm. The authors have developed the 3DcC device along with AJ nanoprinting of the 3D electrodes. The reduced graphene oxide nanoflakes, their functionalization of 3D electrodes, and viral antigens are the critical steps in the development of the 3DcC biosensors device. The 3DcC device was functionalized with the 3D-printed microelectrode by viral antigens using rGO nanoflakes for detection of antibodies from the fluid introduced in the electrochemical cell. The micropillar electrode array is functionalized using rGO nanoflakes, and antigens were bonded with the rGO nanoflakes of the electrode. It was expected that due to the π–π interactions among the coated rGO, the coating was non-uniform [41]. The researchers have analyzed the electrochemical properties of the AJ-printed 3D Au electrode (i.e., with micropillar geometry) by the cyclic voltammetry (CV) method. The experimental results suggest that 3D Au (AJ-printed) and 2D Au (AJ-printed) electrodes exhibit clear oxidation and reduction peaks, where 3D Au electrode showed a 170% enhancement of current compared to the 2D Au electrode. To test the antibody, spike S1 and RBD antigens of SARS-CoV-2 were selected to create the 3DcC testing platform. The sensing performance and impedimetric sensing plots and Rct for the 3DcC sensor are measured under exposure of electrode for phosphate buffer saline, rabbit serum, fetal bovine serum, and spike S1 antibodies (rabbit IgG). This result showed that the graphene oxide sensor was selective to desired specific proteins. The controlled experiments were performed with the 3DcC device with low concentrations of spike S1 antibodies. A clear change in the impedance signal (4.2 kΩ) compared to the sensor baseline and the control serum is detected. Moreover, the device exhibits that beyond 10×10^{-9} m concentration of spike S1 antibodies, the sensor exhibits a saturated impedance signal, due to the maximum number of binding sites on the sensor surface likely to be occupied by the target antibodies.

The discovery of 2D layer–structured materials, such as semiconducting transition metal dichalcogenides, has become promising due to their excellent electrical, biological, chemical, and optoelectronic characteristics. Among them, tungsten disulfide (WS_2) has gained massive attention because of its fascinating properties, such as high carrier mobility, bandgap, very large exciton binding energy, large spin splitting, and polarized light emission, and its other biologically attractive properties. The WS_2 structure consists of three stacked atom layers (S–W–S) bonded together by van der Waals forces. WS_2 nanosheets have shown unparalleled performance in various fields, such as in nanosensors, supercapacitors, energy, hydrogen evolution, catalysts, lithium-ion batteries. Using WS_2 nanosheets, recently, self-signal DNA electrochemical biosensors combined with PIn6COOH have been fabricated by Jimin Yang and his team. They have prepared the WS_2 nanosheets through a simple solvent exfoliation method from bulk WS_2. The PIn6COOH is electropolymerized on the as-grown WS_2 nanosheet–modified carbon paste electrode, and it was used as a platform for electrochemical sensing of the PIK3CA gene from lung cancer. The schematic process of the fabrication process of the self-signal electrochemical-sensing platform using the WS_2 nanosheets is shown in Figure **5.8a–c**. The PBS solution (pH 7.0) containing 1.0×10^{-11} molL⁻¹ probe ssDNA was used and drop-casted to cover the PIn6COOH/WS_2/CPE nanocomposite. The performance of the device was measured by the C–V experiments. Moreover, the whole solution was continually

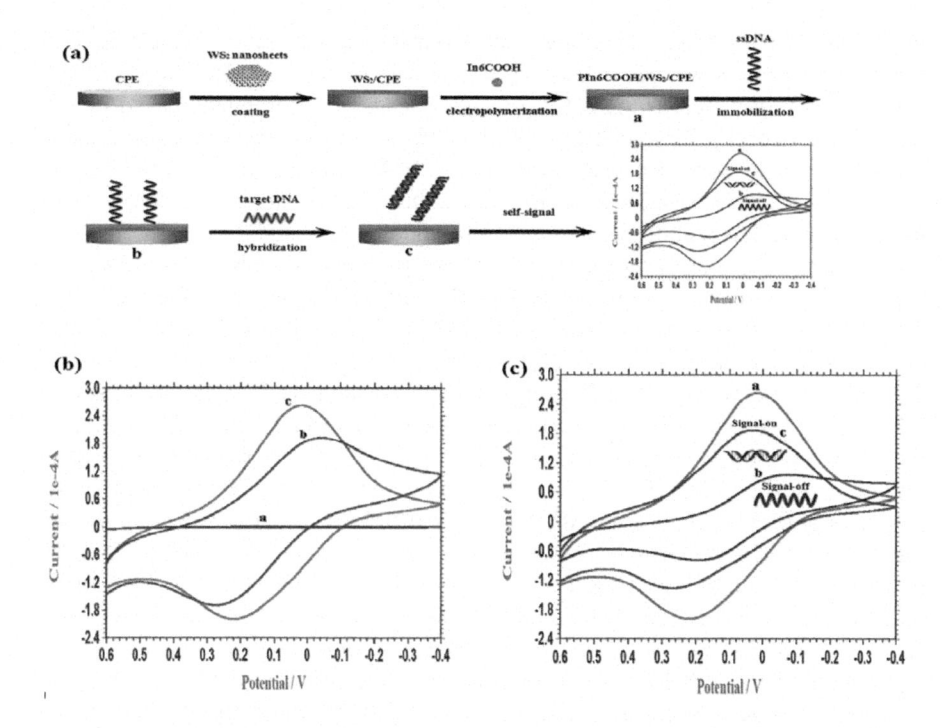

FIGURE 5.8 (a) Schematic process of WS$_2$ nanosheets–based self-signal DNA electrochemical biosensor. (b) C–V result of the bare (i) CPE, (ii) PIn6COOH/CPE, and (iii) PIn6COOH/WS2/CPE in 0.3 mol L_1 PBS (pH 7.0). (c) C–V results of the (i) PIn6COOH/WS$_2$/CPE, (ii) ssDNA/PIn6COOH/WS$_2$/CPE, and (iii) dsDNA/PIn6COOH/WS$_2$/CPE in 0.3 mol L^{-1} PBS (pH 7.0).

bubbled with nitrogen in order to remove the oxygen. The results confirmed that the WS$_2$ nanosheets have significant effect on the performance of the PIn6COOH/WS$_2$ nanocomposite biosensor device. They have experimentally measured that with the increase of WS$_2$ contents, the impedance response of PIn6COOH/WS$_2$ nanocomposite decreased; however, impedance was increased for the amount of WS$_2$ higher than 10 mL of 1.0 mg mL^{-1}.

Moreover, the selectivity of the constructed DNA biosensor shows that when the probe ssDNA was attached on the PIn6COOH/WS$_2$ nanocomposite as noncovalently, the resistance response was increased. When the PIn6COOH/WS$_2$ nanocomposite biosensor was hybridized with the complementary sequence, the resistance response significantly decreased, suggesting that hybridization reaction has occurred, causing the "signal-on." Finally, WS$_2$ nanosheets incorporated with polymer enhances the performance of biosensors in terms of wide dynamic range, high selectivity, and low detection limit.

In addition, compared to MoS$_2$ and WS$_2$ nanosheets, tungsten diselenide (WSe$_2$) based biosensor devices exhibit the lowest detection limits and highest linear-regime sensitivities. Monolayer WSe$_2$ crystals offer higher absorption and excellent surface activity than their multilayers; hence, efficiency of the detection can be greatly improved. Graphene-based FET biosensor device for rapid identification of the SARS-CoV-2 spike protein was previously developed with the limit of detection (LOD) of 1 fg/mL; however, due to the off-state current leakage in graphene-based biosensors, the device also produced false signals. To avoid this issue, two-dimensional WSe_2-based field-effect transistor biosensors for detection of the COVID-19 virus (SARS-CoV-2) have been recently developed. Due to their larger tunable bandgaps, these 2D materials usually decrease the off-state current and give high signal-to-noise ratios. To detect the SARS-CoV-2 spikes proteins, the fabrication process of the 2D-FET biosensors using monolayer WSe_2 crystals is shown in **Figure 5.9a–j**. The 2D material is functionalized with SARSCoV-2 antibody, and for the base-sensing platform, a probe linker of 11- mercaptoundecanoic acid (MUA) is used. Authors have examined the performance of the fabricated 2D-FET sensor for several SARS-CoV-2 spike protein concentrations. The AFM surface topography images of the WSe2 samples before and after the MUA coating process showed a very clean surface with a height profile of ~0.7 nm, while the AFM of the MUA-coated sample exhibits uniform coating of the MUA on the entire WSe_2 crystal with a height profile of 2.8 nm (Figure 5.9a, b). The samples were immersed into the SARS-CoV-2 antibody solution, followed by the addition of the different concentrations of the SARS-CoV-2 spike protein. Electrical transport behavior of the WSe_2 devices with the function of SARS-CoV-2 spike protein concentration was investigated. The I_{Ds}–V_{DS} curves of the WSe_2 device before and after the MUA functionalization were measured (Figure 5.9g–j). The performance was also examined after the attachment of the antibody and spike protein to the FET device, and the I–V characteristics were also measured after each modification process. It was noted that the amplitude and slope of the I_{DS} curve decreased with the functionalization of MUA, which suggests a reduction of the number of mobile charges in the FET device channel. Moreover, it was found that after the introduction of the SARS-CoV-2 antibody and spikes, the device's current amplitude and its slope again slightly increased. A real-time response of the WSe_2-based COVID-19 2D-FET sensor devices with various concentrations of protein spike is demonstrated. The real-time output performance with the function of protein spike concentration is examined by adding 8 µL of the solution drop-wise onto the FET device (Figure 5.9g-j). The I_{DS}-V_{GS} curve shows a clear stepwise change for SARS-CoV-2 antibody functionalized device. The results indicated an effective interaction of SARS-CoV-2 spikes protein with the antibodies. However, the device without the antibody exhibits only a slight change after introducing various spike concentrations. The output results confirmed that for efficient interaction and detection of the SARS-CoV-2 spike protein, SARS-CoV-2 antibody plays an important role in achieving high efficiency of the device system.

5.3 PIEZOELECTRIC BIOSENSORS DEVICE

Piezoelectricity is a well-known physical phenomenon where a non-centrosymmetric material generates a voltage under mechanical deformation, whereas in converse piezoelectric effect, mechanical strain/oscillation is developed under applied electrical field. The crystals without center of symmetry, such as anisotropic crystals, exhibit piezoelectricity [42]. Various materials, such as zinc oxide, quartz (SiO2),

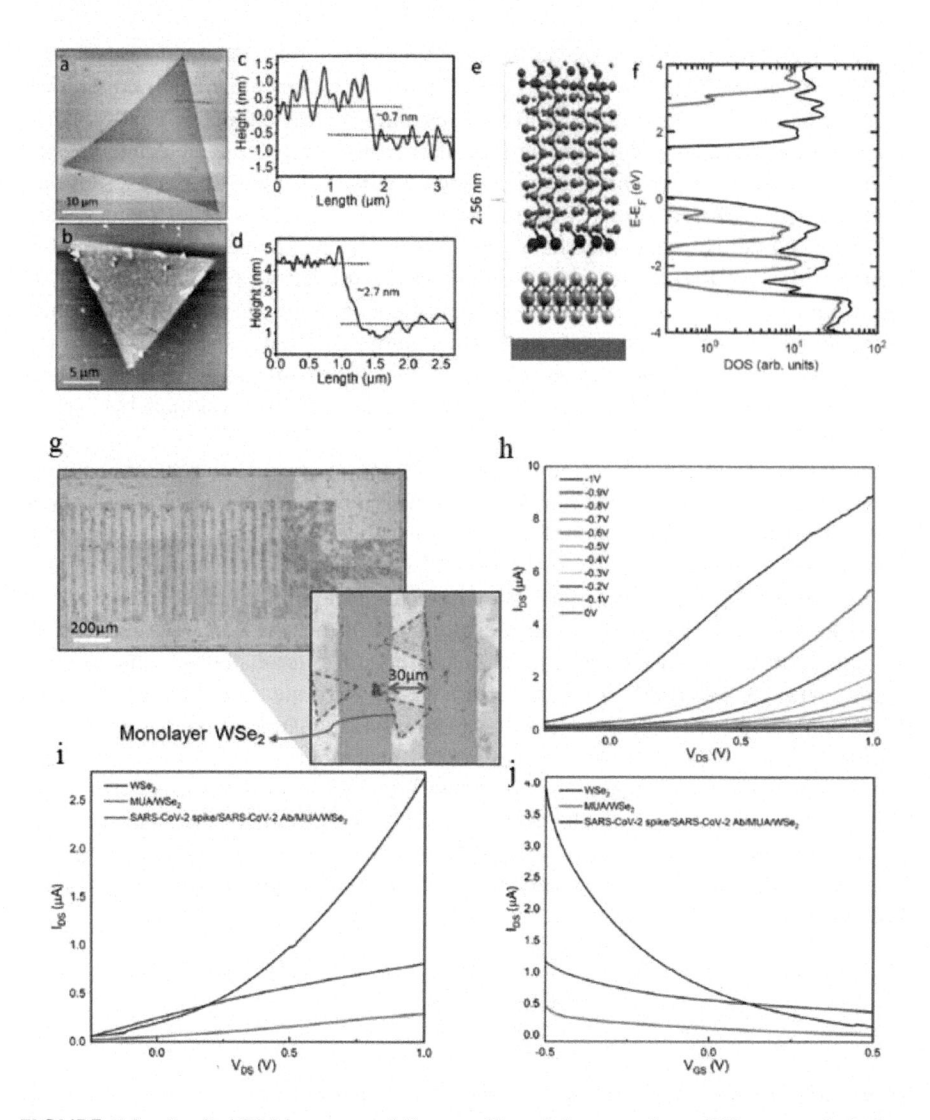

FIGURE 5.9 (a–d) AFM images and line profiles of the monolayer WSe$_2$ crystals before and after MUA functionalization. (e) Schematic image of the MUA monolayer on top of WSe$_2$. (f) The density of states near the Fermi level. (g) Device performance of the WSe$_2$-based biosensors. (h) Optical images of the WSe$_2$ COVID-19 2D-FET sensor. (i–j) Electrical measurement, I$_{DS}$–V$_{DS}$ curves of a pristine WSe$_2$ FET with various gating voltages.

barium titanate (BaTiO$_3$), lead titanate (PbZrTiO$_3$), lithium niobate (LiNbO$_3$), polyvinylidene fluoride (PVDF), polylactic acids (PLA), and peptide materials, exhibit piezoelectric property. The piezoelectric materials have been used for the construction of various biosensors. In 1972, Shons *et al.* developed the piezoelectric immunosensor to detect the cow serum IgG antibody. The piezoelectric biosensor can be excited by alternating voltage, which causes mechanical oscillations of crystal, and the frequency of oscillations is measured [43]. When any other mass (COVID-19 virus, DNA, etc.) was attached on the surface of piezoelectric materials or on the electrodes, the oscillation frequency significantly changed. This variation in oscillation frequency depends on the mass bound on the piezoelectric crystal. Therefore, sensitivity in micrograms, which changes in measurable oscillations, produced the required signal information. Moreover, various other factors, such as thickness, density, and the shear modulus of the piezoelectric materials, also influence the oscillation frequency. Based on the preceding principle, piezoelectric immunosensor biosensors, such as piezoelectric quartz crystal microbalance, have received massive attention due to their high sensitivity, specificity, selectivity, and simplicity. Numerous piezoelectric quartz crystals with high specificity of antigen–antibody immune reaction have recently been developed by various researchers.

Specially, recently, a label-free piezoelectric quartz crystal microbalance immunosensor decorated with gold nanoparticles was developed by Ruchika Chauhan et al. [44]. A self-assembled monolayer of hexandithiol (HDT) and 3D gold nanoparticles have been used to fabricate the piezoelectric biosensor for detection of aflatoxin B1 (AFB1, food mycotoxin). They have used the optimized concentration of 40 g mL^{-1} of aAFB1 antibody for the development of the aAFB1/Cys/AuNPs/HDT/Au immunoelectrode. Electrochemical quartz crystal microbalance cyclic voltammetry technique was used for the detection of AFB1. To measure the response from the biomolecule of the AFB, authors have used the cleaned bare quartz crystal as a reference material. In piezoelectric biosensors, the change in the frequency of oscillation (f) is due to the change in mass (m) deposited on the crystal surface measured. The variation in mass of the electrode or thickness of the deposited material proportionally changes the frequency at which the crystal oscillates. The mathematical relationship between frequency and change in the mass is given by the well-known Sauerbrey formula.

$$\Delta m = -C \frac{\Delta f}{n} \tag{1}$$

Where Δf is the change in frequency (Hz), C is the sensitivity factor of the crystal, and Δm is the change in mass per unit area (g cm^{-2}). The fabrication process of the BSA/aAFB1/Cys/AuNPs/HDT/Au immunosensor and the biochemical interaction between AFB1 and aAFB1 with electrode surface are presented in Figure 5.10a–e. The frequency change is found to be of 8.9 Hz for HDT deposition, whereas mass change was found to be 109.2 ng/cm2. The observed frequency change was 10.16 Hz for the Au nanoparticle deposition, corresponding to 124.66 ng/cm^2 (9.4×10^{14}atoms). It was demonstrated that after functionalization of AuNPs/HDT/Au electrode with cysteamine, frequency change of about 12.7 Hz occurred, which is related to the 8.2×10^{14} number of atoms. It was proposed that increased sensitivity, LOD value, and linear range of the response current on the immunosensor are attributed to increased surface area and high electron transfer rate offered by gold nanoparticles embedded molecularly on the oriented SAM of HDT on Au electrode.

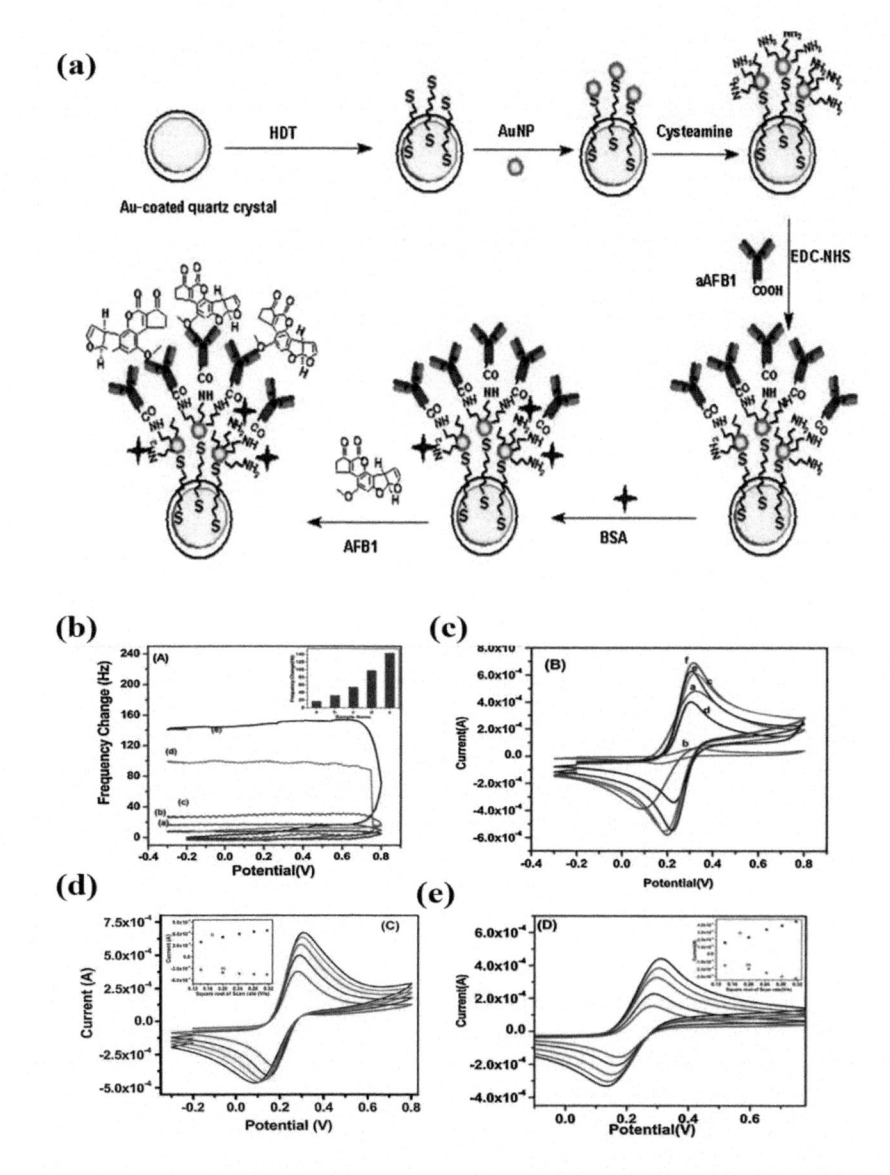

FIGURE 5.10 (a) Schematic diagram of fabrication of immunoelectrode antigen and antibody interaction. (b) The change in frequency of the Au-coated 6 MHz AT-cut quartz crystal. (c) EQCM-CV of (i) bare Au, (ii) HDT/Au, (iii) AuNPs/HDT/Au, (iv) Cys/AuNPs/HDT/Au, and (v) and CV of the Cys/AuNPs/HDT/Au.

The results indicated that the piezoelectric immunosensor exhibits two linear detection ranges: one is from a lower range of 0.008 ngmL^{-1} to 0.3 ng mL^{-1}, and another for the higher range in 1 ng mL^{-1} to 10 ng mL^{-1}, and a sensitivity of 126 μA ng^{-1}mLcm^{-2}. This electrochemical piezoelectric immunosensor is proven as a highly promising technique for detection of AFB1 in real samples.

(a)

(b)

(c)

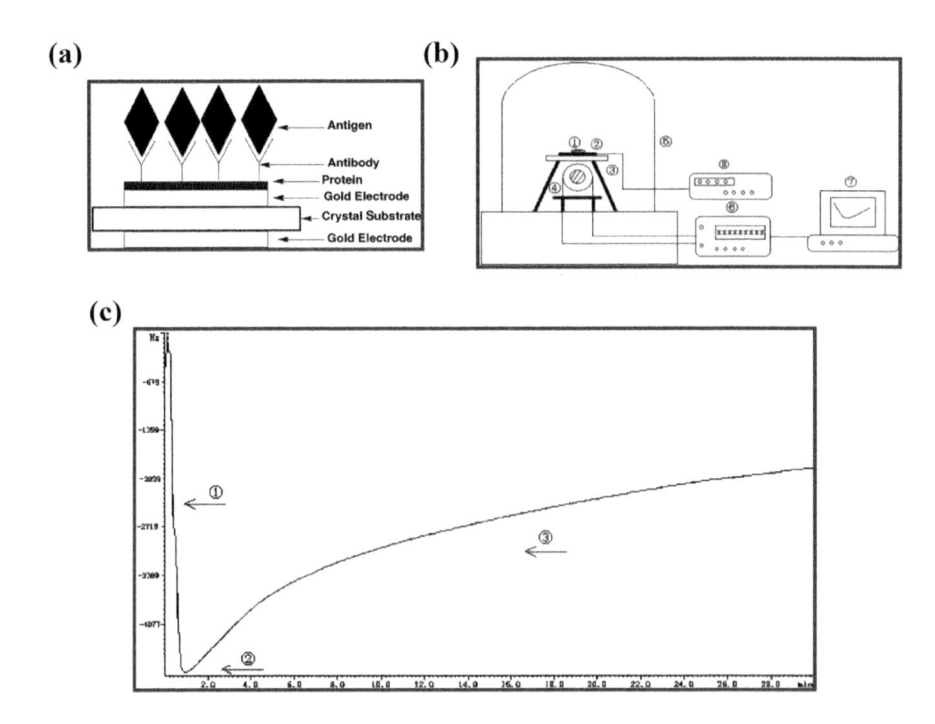

FIGURE 5.11 (a, b) Schematic of adsorption and experimental setup for the measurement of the SARS-COVID virus through piezoelectric biosensors. (c) The response to antigen gas of the piezoelectric crystal (variation of the frequency shift with the elapsed time, where 1 represents the course of adsorption, 2 represents the maximum of frequency shift, and 3 represents the course of desorption).

Moreover, Zuo et al. and his team have developed piezoelectric biosensors for the detection of SARS-associated coronavirus (SARS-CoV) in sputum. To perform the detection of SARS-associated coronavirus, the piezoelectric immunosensor consists of a single crystal with immobilization of an antigen/antibody on the surface used [14].

Due to the specific known mass of the piezoelectric crystal, any changes in the surface mass change its resonant frequency. Based on the earlier equation, the increase of mass at the piezoelectric immunosensor decreases the frequency, and the change in the frequency can be recorded. In the present work, the piezoelectric crystal was coated with the horse polyclonal antibody induced by SARS-CoV, and the protein layer was used to immobilize the antibody. Researchers have reported SARS-CoV aerosol was produced by ultrasonator, and the coated antibody binds with its antigen. Thus, the mass change on the crystal generates a frequency shift that has a linearity relation with the concentration of the antigen (Figure 5.11a–c). It was demonstrated that the frequency shifts have a linear trend with antigen concentration in the range of 0.6–4 µg/mL. The reproducibility of the piezoelectric immunosensor was tested for more than 100 times, and no obvious loss in sensitivity was detected, which confirmed the outstanding stability of the immobilized antibody. Therefore, the studies confirm the potential of piezoelectric biosensors for the efficient detection of the SARS-COVID virus.

REFERENCES

1. Keni R, Alexander A, Nayak PG, Mudgal J, Nandakumar K. COVID-19: emergence, spread, possible treatments, and global burden. Frontiers in Public Health. 2020:216.
2. Nicomedes CJC, Avila RMA. An analysis on the panic during COVID-19 pandemic through an online form. Journal of Affective Disorders. 2020;276:14–22.
3. Mendes T, Carvalho L. Shifting geographies of knowledge production: the coronavirus effect. Tijdschrift voor economische en sociale geografie. 2020;111:205–10.
4. Sharma A, Tiwari S, Deb MK, Marty JL. Severe acute respiratory syndrome coronavirus-2 (SARS-CoV-2): a global pandemic and treatment strategies. International Journal of Antimicrobial Agents. 2020;56:106054.
5. Kevadiya BD, Machhi J, Herskovitz J, Oleynikov MD, Blomberg WR, Bajwa N, Soni D, Das S, Hasan M, Patel M. Diagnostics for SARS-CoV-2 infections. Nature Materials. 2021;20:593–605.
6. Koteswara Rao V. Point of care diagnostic devices for rapid detection of novel coronavirus (SARS-nCoV19) pandemic: a review. Frontiers in Nanotechnology. 2021;22.
7. Ai T, Yang Z, Hou H, Zhan C, Chen C, Lv W, Tao Q, Sun Z, Xia L. Correlation of chest CT and RT-PCR testing for coronavirus disease 2019 (COVID-19) in China: a report of 1014 cases. Radiology. 2020;296:E32–40.
8. Mansour NA, Saleh AI, Badawy M, Ali HA. Accurate detection of Covid-19 patients based on Feature Correlated Naïve Bayes (FCNB) classification strategy. Journal of Ambient Intelligence and Humanized Computing. 2021:1–33.
9. Naresh V, Lee N. A review on biosensors and recent development of nanostructured materials-enabled biosensors. Sensors. 2021;21:1109.
10. Grieshaber D, MacKenzie R, Vörös J, Reimhult E. Electrochemical biosensors-sensor principles and architectures. Sensors. 2008;8:1400–58.
11. Lim JW, Ha D, Lee J, Lee SK, Kim T. Review of micro/nanotechnologies for microbial biosensors. Frontiers in Bioengineering and Biotechnology. 2015;3:61.
12. Hwang MT, Heiranian M, Kim Y, You S, Leem J, Taqieddin A, Faramarzi V, Jing Y, Park I, van der Zande AM. Ultrasensitive detection of nucleic acids using deformed graphene channel field effect biosensors. Nature Communications. 2020;11:1–11.
13. Haleem A, Javaid M, Singh RP, Suman R, Rab S. Biosensors applications in medical field: a brief review. Sensors International. 2021;2:100100.
14. Zuo B, Li S, Guo Z, Zhang J, Chen C. Piezoelectric immunosensor for SARS-associated coronavirus in sputum. Analytical Chemistry. 2004;76:3536–40.
15. Yoon J, Shin M, Lim J, Lee J-Y, Choi J-W. Recent advances in MXene nanocomposite-based biosensors. Biosensors. 2020;10:185.
16. Aykaç A, Gergeroglu H, Beşli B, Akkaş EÖ, Yavaş A, Güler S, Güneş F, Erol M. An overview on recent progress of metal Oxide/Graphene/CNTs-based nanobiosensors. Nanoscale Research Letters. 2021;16:1–19.
17. Nayak P, Anbarasan B, Ramaprabhu S. Fabrication of organophosphorus biosensor using ZnO nanoparticle-decorated carbon nanotube—graphene hybrid composite prepared by a novel green technique. The Journal of Physical Chemistry C. 2013;117:13202–9.
18. Sarkar D, Liu W, Xie X, Anselmo AC, Mitragotri S, Banerjee K. MoS2 field-effect transistor for next-generation label-free biosensors. ACS Nano. 2014;8:3992–4003.
19. Kim S, Ryu H, Tai S, Pedowitz M, Rzasa JR, Pennachio DJ, Hajzus JR, Milton DK, Myers-Ward R, Daniels KM. Real-time ultra-sensitive detection of SARS-CoV-2 by quasi-freestanding epitaxial graphene-based biosensor. Biosensors and Bioelectronics. 2022;197:113803.
20. Pumera M. Graphene in biosensing. Materials Today. 2011;14:308–15.
21. Zhu Z. An overview of carbon nanotubes and graphene for biosensing applications. Nano-Micro Letters. 2017;9:1–24.

22. Armano A, Agnello S. Two-dimensional carbon: a review of synthesis methods, and electronic, optical, and vibrational properties of single-layer graphene. C: Journal of Carbon Research. 2019;5:67.
23. Ohno Y, Maehashi K, Matsumoto K. Label-free biosensors based on aptamer-modified graphene field-effect transistors. Journal of the American Chemical Society. 2010;132: 18012–3.
24. Chen T-Y, Loan PTK, Hsu C-L, Lee Y-H, Wang JT-W, Wei K-H, Lin C-T, Li L-J. Label-free detection of DNA hybridization using transistors based on CVD grown graphene. Biosensors and Bioelectronics. 2013;41:103–9.
25. Smith AT, LaChance AM, Zeng S, Liu B, Sun L. Synthesis, properties, and applications of graphene oxide/reduced graphene oxide and their nanocomposites. Nano Materials Science. 2019;1:31–47.
26. Tiwari SK, Sahoo S, Wang N, Huczko A. Graphene research and their outputs: status and prospect. Journal of Science: Advanced Materials and Devices. 2020;5:10–29.
27. Novodchuk I, Bajcsy M, Yavuz M. Graphene-based field effect transistor biosensors for breast cancer detection: a review on biosensing strategies. Carbon. 2021;172:431–53.
28. Thanh TD, Chuong ND, Van Hien H, Kshetri T, Kim NH, Lee JH. Recent advances in two-dimensional transition metal dichalcogenides-graphene heterostructured materials for electrochemical applications. Progress in Materials Science. 2018;96:51–85.
29. Mathew M, Radhakrishnan S, Vaidyanathan A, Chakraborty B, Rout CS. Flexible and wearable electrochemical biosensors based on two-dimensional materials: recent developments. Analytical and Bioanalytical Chemistry. 2021;413:727–62.
30. Mitta SB, Choi MS, Nipane A, Ali F, Kim C, Teherani JT, Hone J, Yoo WJ. Electrical characterization of 2D materials-based field-effect transistors. 2D Materials. 2020;8:012002.
31. Kajale SN, Yadav S, Cai Y, Joy B, Sarkar D. 2D material based field effect transistors and nanoelectromechanical systems for sensing applications. Iscience. 2021;24:103513.
32. Sharma BK, Ahn J-H. Graphene based field effect transistors: efforts made towards flexible electronics. Solid-State Electronics. 2013;89:177–88.
33. Seo G, Lee G, Kim MJ, Baek S-H, Choi M, Ku KB, Lee C-S, Jun S, Park D, Kim HG. Rapid detection of COVID-19 causative virus (SARS-CoV-2) in human nasopharyngeal swab specimens using field-effect transistor-based biosensor. ACS Nano. 2020;14:5135–42.
34. Béraud A, Sauvage M, Bazán CM, Tie M, Bencherif A, Bouilly D. Graphene field-effect transistors as bioanalytical sensors: design, operation and performance. Analyst. 2021;146:403–28.
35. Seifi T, Kamali AR. Antiviral performance of graphene-based materials with emphasis on COVID-19: a review. Medicine in Drug Discovery. 2021;11:100099.
36. Yang J, Gao L, Peng C, Zhang W. Construction of self-signal DNA electrochemical biosensor employing WS 2 nanosheets combined with Pin6COOH. RSC Advances. 2019;9:9613–9.
37. Wang X, Zhu Y, Olsen TR, Sun N, Zhang W, Pei R, Lin Q. A graphene aptasensor for biomarker detection in human serum. Electrochimica Acta. 2018;290:356–63.
38. Papamichael KI, Kreuzer MP, Guilbault GG. Viability of allergy (IgE) detection using an alternative aptamer receptor and electrochemical means. Sensors and Actuators B: Chemical. 2007;121:178–86.
39. Mukherjee S, Meshik X, Choi M, Farid S, Datta D, Lan Y, Poduri S, Sarkar K, Baterdene U, Huang C-E. A graphene and aptamer based liquid gated FET-like electrochemical biosensor to detect adenosine triphosphate. IEEE Transactions on Nanobioscience. 2015;14:967–72.
40. Liu X, Liang R, Gao G, Pan C, Jiang C, Xu Q, Luo J, Zou X, Yang Z, Liao L. MoS2 negative-capacitance field-effect transistors with subthreshold swing below the physics limit. Advanced Materials. 2018;30:1800932.

41. Ali MA, Hu C, Jahan S, Yuan B, Saleh MS, Ju E, Gao SJ, Panat R. Sensing of COVID-19 antibodies in seconds via Aerosol Jet nanoprinted reduced-graphene-oxide-coated 3D electrodes. Advanced Materials. 2021;33:2006647.
42. Aktas O, Kangama M, Linyu G, Catalan G, Ding X, Zunger A, Salje EK. Piezoelectricity in nominally centrosymmetric phases. Physical Review Research. 2021;3:043221.
43. Pohanka M. Overview of piezoelectric biosensors, immunosensors and DNA sensors and their applications. Materials. 2018;11:448.
44. Chauhan R, Singh J, Solanki PR, Manaka T, Iwamoto M, Basu T, Malhotra B. Label-free piezoelectric immunosensor decorated with gold nanoparticles: kinetic analysis and biosensing application. Sensors and Actuators B: Chemical. 2016;222:804–14.

6 3D Printing to Design and Develop the Components for COVID-19

Gaurav Kumar Gupta

3D printing is an additive manufacturing (AM) technique for producing a variety of components, structures, and complex geometries from CAD data [1]. The schematic of 3D printing is shown in Figure 6.1. The process involves deposition of materials layer by layer. 3D printing technique was first time used by the Charles Hull in 1986 [2]. The process invented initially was stereolithography (SLA). Later on, new techniques were evolved for 3D printing of various polymers, metals, ceramics, and composites. Additive manufacturing (AM) has been used in various industries that include biomedical, aerospace, general engineering, defense, jewelry construction, design, etc. The developments in this area have reduced the cost of 3D printers. 3D printing has fast and cost-effective prototyping capability that has reduced the time and cost for developing a product. By using 3D printing technique, we can produce small quantities of customized product. Currently, customized product by 3D printing is almost 50% of the total 3D printing components [3]. Custom-built components by 3D printing are most pronounced in the biomedical field as patient-specific medical implants can be printed from CT data [4].

The advantages of 3D printing over conventional methods are the production of components with complicated geometry, high precision, and less material wastage. Additive manufacturing is capable of producing parts of various sizes from the micron to the meter scale. However, the precision is dependent upon the printing method and the scale of printing. AM also has the capability of producing complex geometries, such

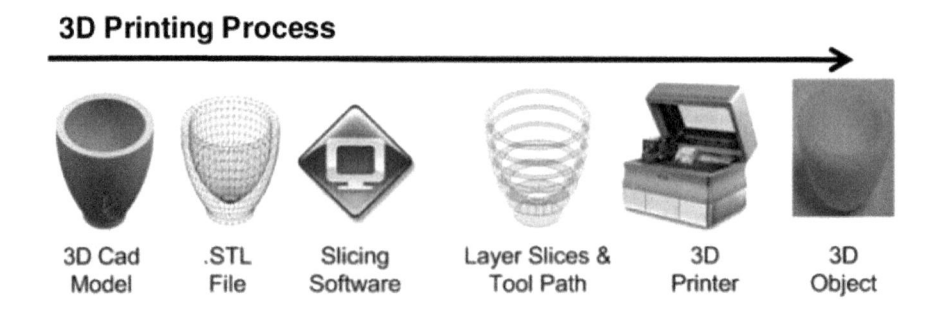

FIGURE 6.1 Steps involved in the 3D printing of a desired object.

Source: Reproduced from ref. [1].

  DOI: 10.1201/9781003316381-6

as lattice structures and porous components, wherein conventional methods are time-consuming and require tooling and post-processing. Some of the critical issues in 3D printing are resolution, surface finish, and interlayer bonding. Sometimes, an inferior mechanical property due to defects and anisotropy in properties in 3D-printed parts limits the sizes of the components. Therefore, research is required for enhancing the mechanical properties, reducing defects, and less anisotropy. Futuristic approach in 3D printing is to use computer-aided design with advanced simulation capabilities. There is also a requirement in improvement of 3D printers that can print with high speed and can reduce the cost. The following sections review the various techniques of 3D printing.

6.1 DIFFERENT METHODS OF ADDITIVE MANUFACTURING

Various methods of additive manufacturing (AM) have been developed depending on resolution, component complexity, and material to be printed. Every method aims to print complex structures with high resolution and less defects and improved mechanical properties. The most versatile technique of 3D printing is fused deposition modeling (FDM), also known as material extrusion process, which uses polymer filaments/wires for printing. The other 3D printing methods are selective laser sintering (SLS) or powder bed fusion, selective laser melting (SLM), stereolithography (SLA) or vat polymerization, inkjet printing (material jetting), binder jetting, direct energy deposition (DED), and laminated object manufacturing (LOM) or sheet lamination [5]. All AM processes are shown in Figure 6.2. These methods are explained

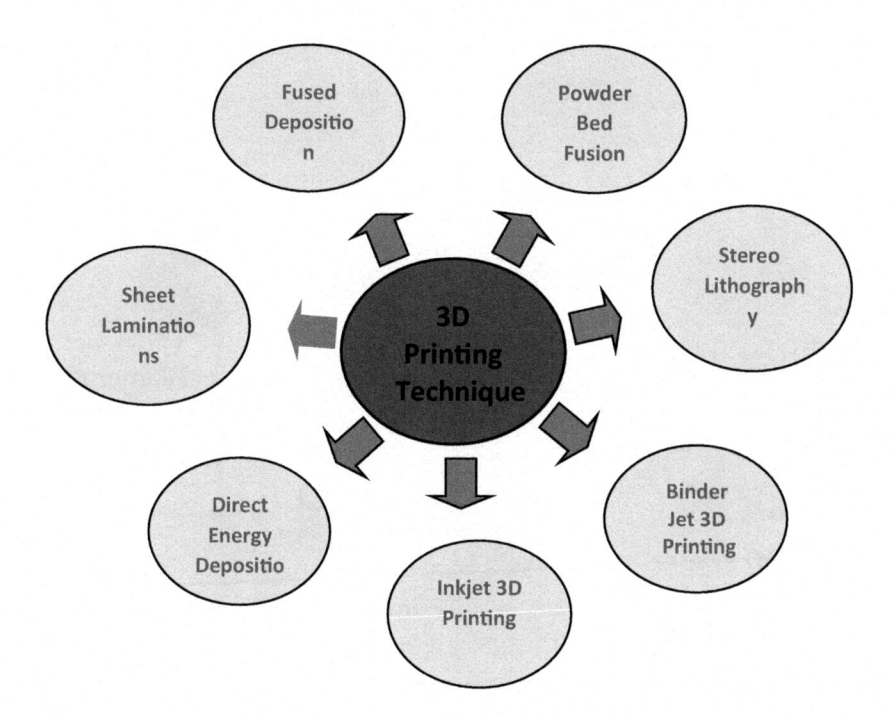

FIGURE 6.2 Different methods of 3D printing.

in subsequent sections. In these sections, the applications and suitable materials for each technique are discussed along with the benefits and drawbacks. Some novel techniques have been developed in recent years, such as two-photon polymerization (TPP) [6], projection micro-stereolithography (PµSLA) [7], and electrochemical additive manufacturing (ECAM) [8].

6.2   FUSED DEPOSITION MODELING (FDM)

The FDM printing method has lower cost, is very simple, and has a very high speed of printing. In this method, a thermoplastic polymer filament is used to print different layers of solid model, thus creating a 3D model (Figure 6.3) [9]. The filament is converted into a viscous liquid by passing it through a heater. The heater is located at the nozzle, and later, the viscous material is extruded through the nozzle and deposited over the printing bed and then solidifies. Thus, the new layer gets fused with the previously formed layer. The layer thickness, printing speed, and printing strategy are the important processing parameters that control the final properties of the component [10]. Many a times, less fusion between layers, interlayer distortion, and poor surface finish occur in components, and as a result, lower mechanical properties are obtained [11]. Currently, fiber-reinforced composites are also manufactured using the FDM process, as the fiber reinforcing gives more strength to 3D-printed components [12]. For this kind of printing, fiber- or particle-reinforced thermoplastic filaments are used [12]. During the printing of fiber-reinforced plastic, random fiber orientation, particle agglomeration, poor interfacial bonding between the fiber/particle, and matrix and porosity/cracks formation are some of the challenges that need to be solved [13].

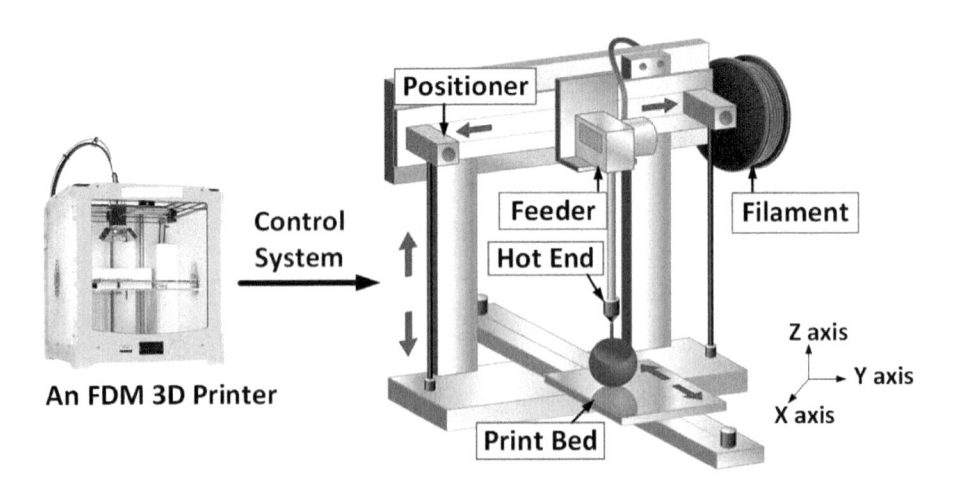

FIGURE 6.3   A schematic of an FDM 3D printer.

Source: Reproduced from ref. [9].

6.3 POWDER BED FUSION

In the powder bed fusion process, very fine powders are used. These powders are spread in the form of closely packed layer on a printing platform. These powders are fused together with a focused laser light. The focused laser is scanned over the printing platform according to 3D model design layer by layer. After complete laser powder interaction on a layer, another layer of powder is spread on top of the previously fused layer. This process continues until the final 3D part is built. The remaining powder which is not fused is then removed by vacuum or manually. Powder size distribution and shape, its bulk density, laser absorption, and particle chemical composition are the most common factors that determine the density and mechanical properties of the printed part [14]. The laser interaction with the powder can be employed in two processes: selective laser sintering (SLS) and selective laser melting (SLM). The powder bed process is shown in Figure 6.4 [15]. SLS is used for sintering of high-melting-point powders and melting of low-melting-point powders. SLS is normally used for polymers, polymer composites, and few metals and alloys, whereas SLM can be used for most metals and alloys. In case of SLS, laser interaction with powder doesn't melt the powders and rather elevates the surface temperature of the powder enough to allow diffusion at atomic level, thus allowing powders to fuse together [16], whereas in the case of SLM, powders are melted fully and form new surfaces [16]. The mechanism of both processes is shown in Figure 6.5. Components made by the SLM route have higher density, smooth surface finish, and superior mechanical properties [17]. Laser power density, scanning speed, and scan path trajectory are the main parameters of the powder bed process. In comparison to FDM, higher resolution and high surface finish are some of the benefits of powder bed fusion. SLS/SLM process is used in many industries for the production of complex components for various applications, such as scaffolds and patient-specific implants for biomedical industries, lattices, and turbine blades, engine components for aerospace, and automotive sectors [17]. Some of the limitations of the powder bed processes are very high machine and raw materials cost and higher printing time.

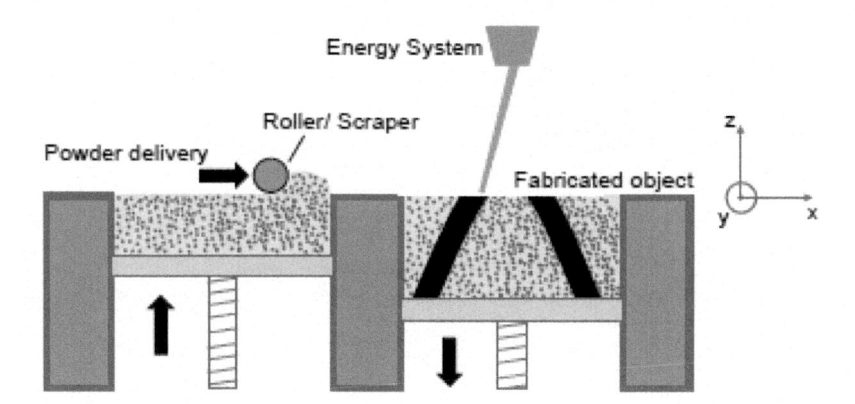

FIGURE 6.4 Powder bed process (SLS and SLM).

Source: Reproduced from ref. [15].

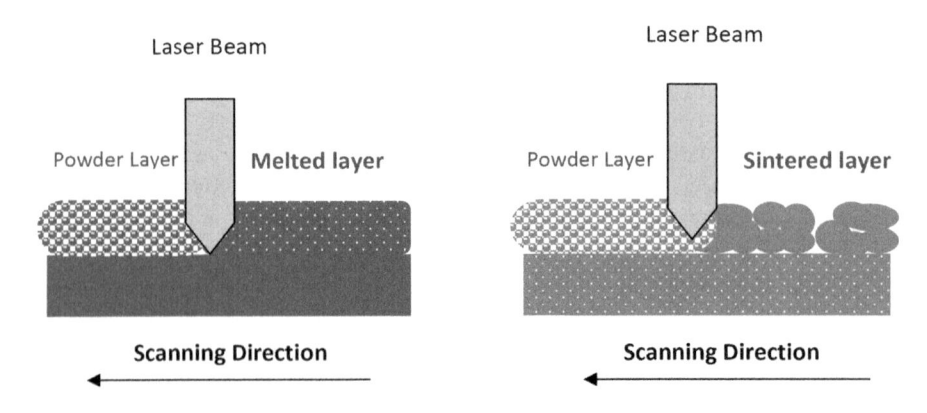

FIGURE 6.5   Process differentiation between SLS and SLM.

6.4   INKJET 3D PRINTING

Inkjet printing or material jetting is a widely used process for 3D printing of ceramics. It is mostly used for the synthesis of complex ceramic components, for example, scaffolds and porous ceramics. In inkjet printing, the ceramic suspension is pressurized and passes through an injection nozzle in the form of fine droplets. These droplets solidify on the substrate and form a continuous structure. The previously printed layers are adhered with new layers. Further, the whole component is to be sintered in order to have higher strength due to the formation of chemical bonding. The schematic of the inkjet process is shown in Figure 6.6 [18].

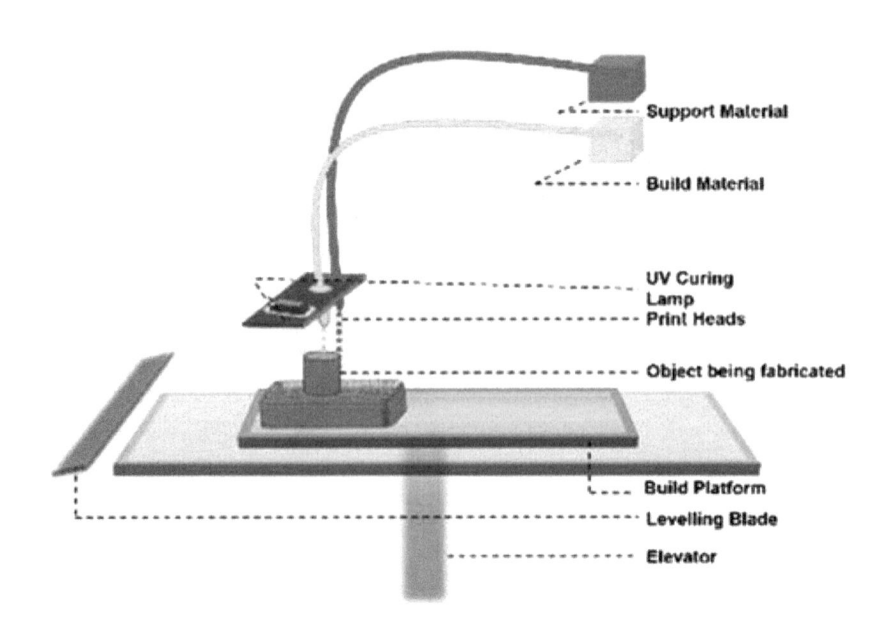

FIGURE 6.6   Schematic of inkjet 3D printing.

Source: Reproduced from ref. [18].

The inkjet method is very fast, flexible, and highly efficient. One can print complex structures. Two types of suspension are used. One is wax-based, and another is aqueous. Wax-based suspensions are used in melted form and solidify while printing on cold substrates, whereas liquid suspensions are deposited on substrates and solidified by evaporation [19]. The factors that are important in inkjet printing are size distribution and shape of powder, viscosity of the suspension, printing speed, and nozzle diameter [20]. Some of the limitations of inkjet printing are workability of suspension, poor surface resolution, and lower bonding between subsequent layers.

6.5 STEREOLITHOGRAPHY (SLA)

SLA is one of the primitive methods of 3D printing developed in 1986 [2]. In this method, UV light is used to cure or polymerize the resin. These monomers are basically acrylic or epoxy-based and are UV-curable. UV interaction with resin polymerizes the monomer chains. Later, the polymerized resin layer in the printing head is solidified for further fusion of subsequent resin layers. The unreacted resin is taken out from the container after printing. A schematic of SLA machine is shown in Figure 6.7 [21]. Sometimes, post-processing, such as heating or curing, is used for some components to get the higher mechanical properties. SLA printing can produce very high-quality parts with resolution as low as 10 μm [13]. The SLA process is relatively slow and expensive. A very small range of materials is suitable for SLA.

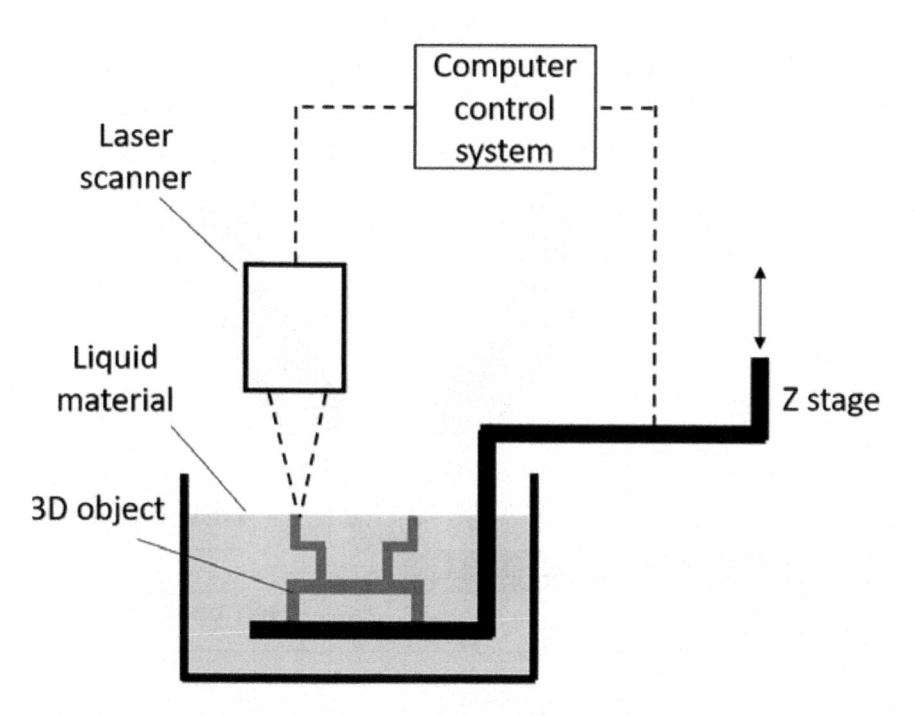

FIGURE 6.7 A schematic of SLA (stereolithography) process.

Source: Reproduced from ref. [21].

The main factors of SLA printing are wavelength and energy of UV source, exposure time, and viscosity of resin [22]. SLA has also been used in fabrication of polymer nanocomposites–based components [23]. A dispersion of ceramic particles in monomers can be used to print ceramic–polymer composites [18]

6.6   DIRECT ENERGY DEPOSITION (DED)

DED or direct energy deposition is also known as laser-engineered net shaping (LENSTM), directed light fabrication (DLF), or direct metal deposition (DMD) [24]. This process uses the laser or electron beam as source of energy. In this process, the beam is directly focused to melt the input material in the form of powder or wire. The same beam is also used to melt the desired small region in the substrate. The input liquid melt is then deposited onto the substrate and fused with the substrate during solidification [23]. The DED process is shown in Figure 6.8 [25].

In the DED process, no powder bed is used, and the input material is melted before deposition. The DED process is useful in cracks filling and modification in manufactured part. Using DED, multiple-axis deposition and multiple input materials are possible [26]. The DED technique is most commonly used in aerospace application, wherein alloys used are titanium, superalloys, stainless steel, etc. This method is currently being used in repairing turbine engines [26]. DED-printed components have excellent mechanical properties, refined microstructure, and tight composition

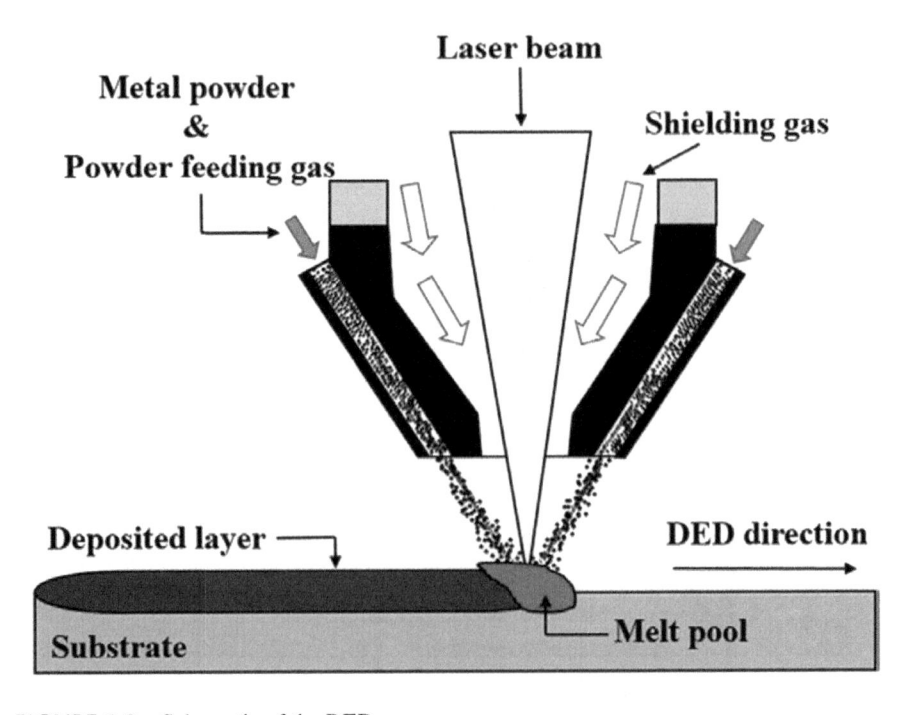

FIGURE 6.8   Schematic of the DED process.

Source: Reproduced from ref. [25].

control. Moreover, the DED process takes less time and is cost-effective. In spite of having various advantages, the DED process has lower resolution and surface roughness. Less-complex parts are possible with this method [24].

6.7 LAMINATED OBJECT MANUFACTURING (LOM)

Laminated object manufacturing (LOM) is a subtractive-additive processing technique. In this process, layer-by-layer cutting is conducted initially, and joining of the lamination of sheets is carried out. Material layers are cut by CNC-based laser sheet cutter, and cut lamination is then bonded together using heat [27]. The materials left over after completion of the process can be removed or recycled [28]. LOM is suitable for various materials, such as polymer, ceramics, paper, and metal sheets. Sometimes, post-processing is required to enhance the mechanical properties. Ultrasonic additive manufacturing is a type of LOM process. The schematic of the LOM process, especially ultrasonic additive manufacturing, is shown in Figure 6.9 [29]. In this process, ultrasonic materials joining along with and CNC milling are used simultaneously [30]. LOM is now being used in many industries, such as in metals, paper, electronics, and construction.

The LOM process is used to make smart structures in construction and general engineering applications. LOM can create cavities in the structure for further addition of devices and sensors. LOM is one of the best 3D printing methods for larger components and structures. In spite of these advantages, LOM has limitations in terms of poor surface finish, lower resolution, and time-taking process. It is also not suitable for complex-shaped components.

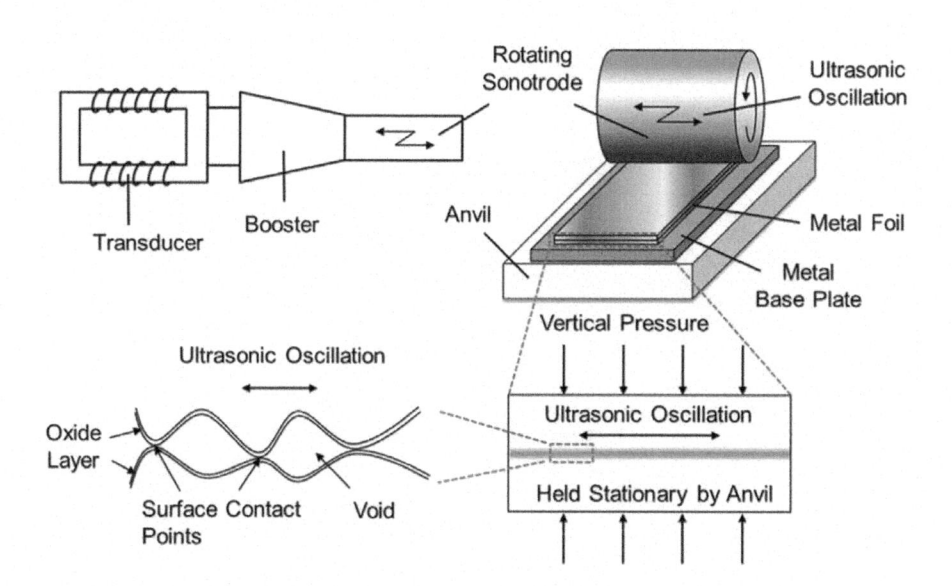

FIGURE 6.9 Ultrasonic additive manufacturing process.

Source: Reproduced from ref. [29].

6.8　BINDER JET 3D PRINTING

Binder jet 3D printing is a 3D printing process in which a liquid binder is deposited over powder layers. This process is similar to powder bed process, wherein the laser is scanned over the desired geometry, but here the binder is deposited in the same fashion as a laser [31]. This liquid binder comes from the nozzle in the form of a jet. The areas where the binder is poured are selectively glued with each other, with the remaining area unglued. After completion of binder jetting in all the layers, the components undergo de-binding and sintering in order to remove the binder and densification of the components, respectively [31]. The schematic of the binder jet 3D printing is shown is Figure 6.10 [32]. Among all AM technologies, this process is quite promising due to fast production of complex components [33]. In this process, powder metallurgy knowledge is required, and components can be produced with similar material properties and surface finish produced by traditional powder metallurgy processing. Many materials have been 3D-printed using this process, but still, knowledge about densification after sintering and post-heat treatment in many material systems are yet lacking [34]. The important factors that affect the binder jetting process on properties are powder shape and size distribution, printing parameters (layer thickness, print orientation, binder concentration, print speed, and drying time), and sintering cycle.

6.9　3D PRINTING FOR BIOMEDICAL INDUSTRY

The biomedical applications of 3D printing are about 10% of the total 3D printing market, and it is continuously growing with time. Using 3D printing, biomedical implants, engineered organs, drug delivery systems, and medical industry machine components can be produced [35]. Extremely complex components can be produced by a variety of materials ranging from polymer to metals, ceramic, and their composites. The 3D printing process is suitable in printing patient-specific biomedical

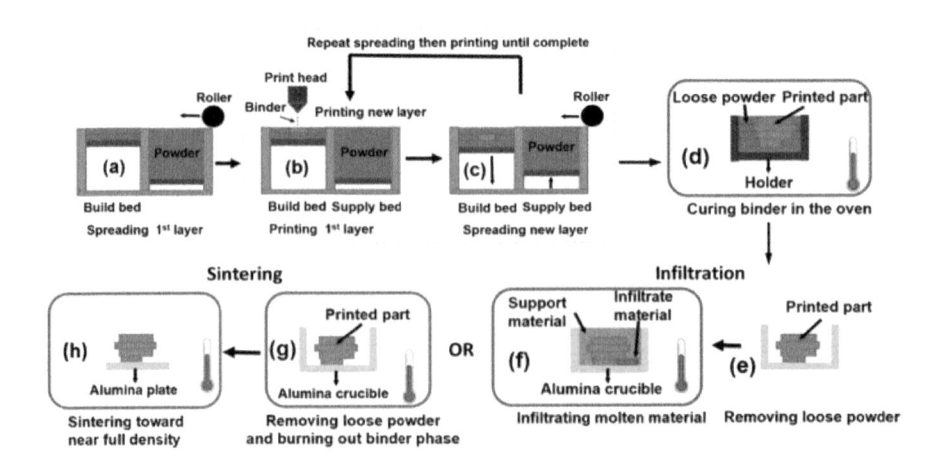

FIGURE 6.10　Binder jetting printing process.

Source: Reproduced from ref. [32].

components, for example, hearing, orthopedic implants, and surgical components, and personal protective equipment for clinical atmosphere [36, 37].

3D printing is more cost-effective for lower production volumes as well as doesn't require new tooling fixtures each time for newer components [38]. One of great advantages in 3D printing is the sharing of CAD model of the components in very short time so that anyone can print similar components in their own premises. One of the key requirements in bio 3D printing is biocompatibility so it does not trigger immune and inflammatory responses [39]. Therefore, there is need to print prototype components that can be used for toxicity tests, disease models for testing adverse reactions to drugs [40]. Using 3D printing, geometrical features like lattice structures can be added in implants. The lattice structure reduces stress-shielding between the implant and the bone [41]. Graded Lattices structures can also be printed using 3D printing [42]. Titanium alloys and biodegradable materials (polymer and metal) lattice are being produced in biomedical industry. The implants are also tested for in vitro and in vivo conditions. Research is also going on producing 3D-printed Mg stents and screws-based biodegradable implants, as they have optimized mechanical and degradation response without body-adverse response [43].

In all biomedical implants, there are always some regulatory issues. All 3D-printed biomedical components require FDA approval [44]. Many of the patient-specific 3D-printed implants are very easy to manufacture but require rigorous trials before clinical use [45]. Moreover, important factors which are checked during trials are mechanical and physical properties: cell viability, nutrients transport, antibiotics response, biocompatibility, and bio-absorbability [46]. Currently, devices and implants are being developed to improve aesthetics, functionality, and comfort. Studies are being conducted to print devices that are easy to adapt and manufacture at lower cost so that countries with inadequate and poor health care can also take advantages of the same. In order to get maximum output using 3D printing in biomedical industry, some challenges are to be overcome. These are limited materials for 3D printing, lower biocompatibility for best materials, inconsistency in properties, etc. [47]. Thus, newer process and materials should be developed with optimized processing parameters. In spite of the many challenges, future research and development would definitely help upscale implants and devices for clinical applications with cost-effectiveness.

6.10 NEED OF 3D PRINTING FOR COVID MANAGEMENT

The novel coronavirus disease (COVID-19) pandemic has affected more than 200 countries around the globe [48]. This disease is normally transmitted through respiratory droplets during direct contact with an infected person or touching contaminated objects and surfaces [49]. Due to the contagious nature of the virus, the need for medical and specialized products, especially personal protective equipment (PPE), was enhanced manifold. This has resulted into critical shortages of essential products. During that time, WHO directed industries and governments to increase the supplies to meet demands. Since the conventional production capacity is not able to supply against the excessive global demand, manufacturers have therefore referred to 3D printing to fill the gap and increase the production line of medical devices. In this context, conventional designs have been modified to suit the 3D printing requirements. The most important components were PPEs and other medical supplies, gloves, goggles, and face shields were initially the choice for 3D printing [50]. In this regard, advanced manufacturing networks along with 3D printing factories were established. These 3D printing factories were set up near hospitals to quickly serve the needs of the medical industry. In the future also, 3D printing can help the world in critical events. In further sections, several 3D-printed components produced by various companies to fight the coronavirus are discussed.

6.11   3D-PRINTED COMPONENTS

The 3D-printed components that were used in the early days of the pandemic were hands-free handle, face masks, ventilator valves, face shields, and swabs. The broad spectrum of 3D-printed components for COVID-19 control and management includes personal protective equipment [51], medical and testing devices [52], personal accessories, visualization aids, and emergency dwellings [53]. The broad spectrum of 3D-printed components for COVID control is shown in Figure 6.11. Some of the 3D-printed devices are shown in Figure 6.12 [54]. Most of these components are based on polymers and polymer-based composites. Hence, the 3D printing processes mostly used for these components were FDM, SLA, and SLS. The development of these 3D printing devices is discussed separately in the following sections.

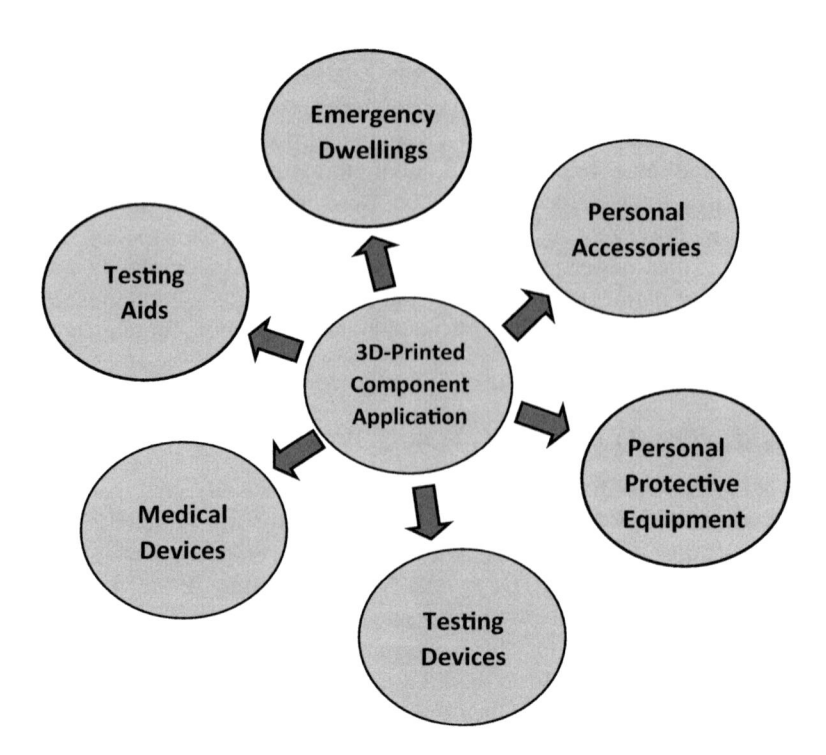

FIGURE 6.11   3D-printed applications for COVID management and control.

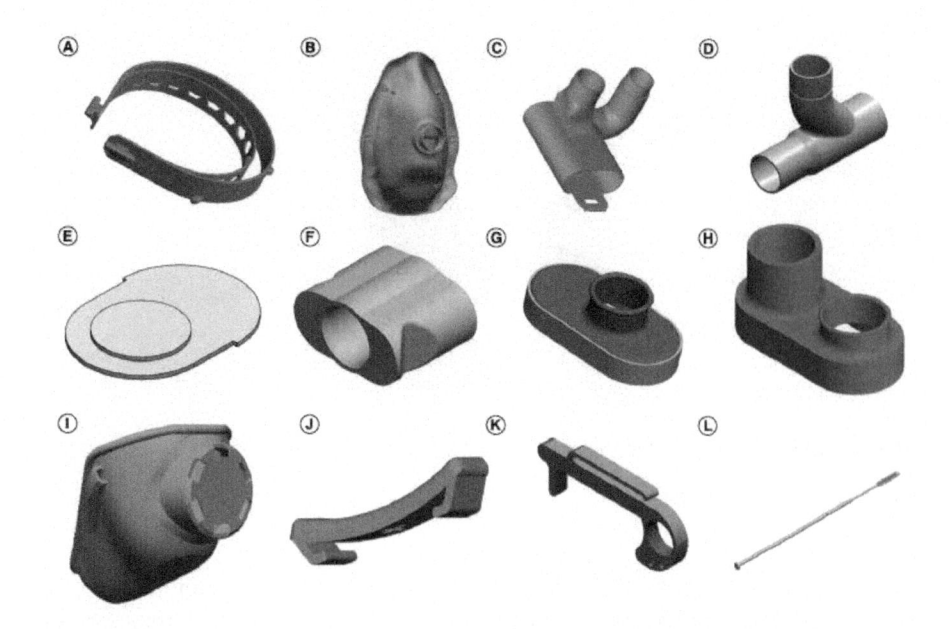

FIGURE 6.12 Devices for COVID-19 pandemic situation: (a) face shield holder, (b) mask, (c) adaptor to ventilator, (d) ventilator adaptor, (e) face shields, (f) N95 filter assembly, (g) valve, (h) body valve, (i) whole mask, (j) mask-holding clip, (k) door handle, and (l) swab.

Source: Reproduced with permission from ref. [54].

6.12 MEDICAL DEVICES

During the time of COVID, ventilator machines shortage was there. Hence, continuous positive airways pressure (CPAP) has been used for critical patients. A 3D-printed valve known as Charlotte valve [55] was designed by the Isinnova Company. This valve works as a mask connector to CPAP machines. As ventilators were fewer, doctors also have used splitting devices for single ventilators for parallel supply of oxygen to multiple patients. Formlabs and other companies designed and produced tubing splitters suitable for the parallel use of a single oxygen supply from ventilators [56]. Elsewhere 3D Systems in the UK produced 2,880 ventilator valves with nylon by using selective laser sintering (SLS) using the SLS ProX 6100 [57]. The firm also produced splitters for ventilators that have saved the lives of nearly 2.4% of patients [57].

6.13 PERSONAL PROTECTIVE EQUIPMENT (PPES)

During the pandemic, face shields were the popular prevention solutions to provide a barrier between the person and surrounding contacts. Many 3D printing companies, like Formlabs, Prusa, and Stratasys, have not only produced face shield but also provided the 3D model design in open source [56, 58]. In spite of good protection due to 3D-printed face shield, there is a chance that the eyes are not protected, and the

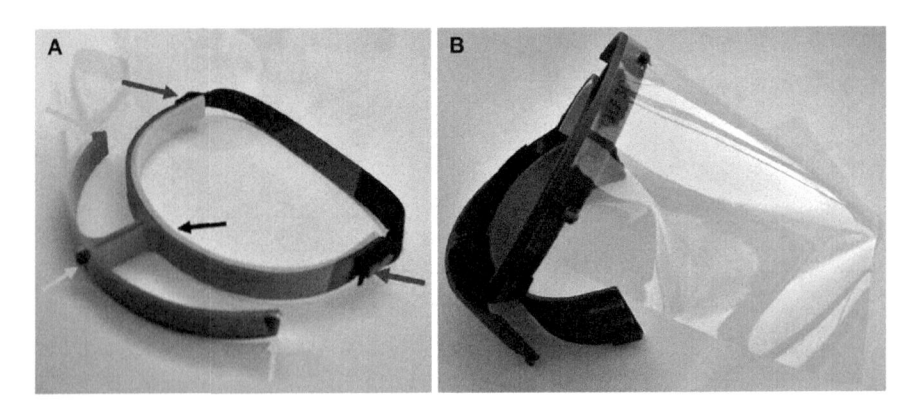

FIGURE 6.13   3D-printed face shield.

Source: Reproduced from ref. [60].

virus can be transmitted through the eyes. In this regard, WHO recommended the use of goggles and face shields simultaneously [59]. Some companies have produced bigger-size face shields to accommodate goggles too [60]. One of the designs of 3D-printed face shield is shown in Figure 6.13.

6.14   TESTING DEVICES

During SARS-CoV-2 testing, a nasal or mouth swab is used to collect nasal/mouth secretions. 3D printing is also one of the options to produce nasopharyngeal swabs. Using 3D printing, complex tip structures in the swabs can be created for enhanced collection efficacy. Different companies and institutes, like Formlabs, Harvard University, Massachusetts Institute of Technology, etc., have designed and fabricated nasal swabs using 3D-printed materials [61]. 3D Systems has produced 3D-printed swabs from FLEX-CLR 20 [57]. The printed part is biocompatible, can be auto-claved, and can bend at least 180 degrees without breaking. Some of the 3D-printed swabs are shown along with conventional swabs in Figure 6.14 [62].

6.15   PERSONAL ACCESSORIES

A shortage of PPE was created during the COVID-19 pandemic, which had created panic among frontline health workers and support staff. In this situation, 3D-printed face masks provided a great relief. In contrast to disposable PPE kits, which are responsible for medical waste accumulation, 3D printing provides solutions to conserve natural resources by using recyclable materials and reusing filters. 3D Systems developed the emergency Stopgap Face Mask [57]. The component was printed from nylon material (Duraform PA or Duraform ProX PA) using selective laser sintering technology. The face mask is comprised of the 3D-printed mask and filter cover, two elastic strips, and a rectangular patch of filter material. This mask can be cleaned

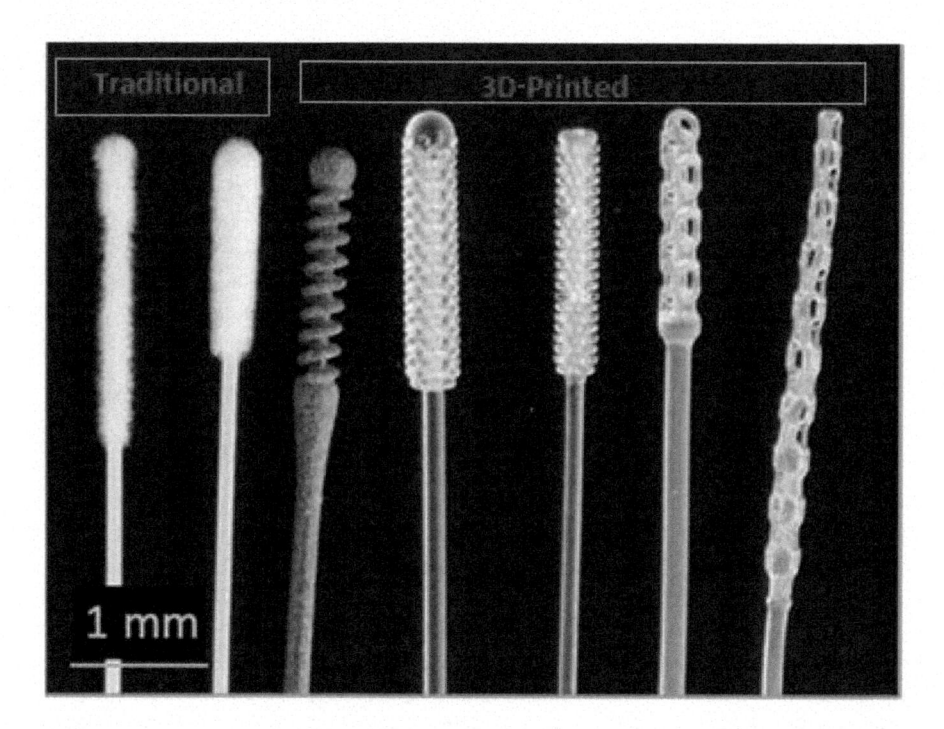

FIGURE 6.14   Traditional swabs vs. 3D-printed swabs.

Source: Reproduced from ref. [62].

using disinfectants or by autoclave. One more personal component is the hands-free door handle attachment that allows opening doors using the elbows/feet without any direct contact with the door handle [63]. This door handle can also be 3D-printed.

6.16   TESTING AIDS

In order to give health workers more support and training, sometimes, organ models are used that are called manikins. Complex organ models can be produced by 3D printing. Facial organ manikin is thus printed with transparent materials for learning treatment for the nasal cavity, throat, and mouth. Using these manikins, healthcare workers practiced swab collections. In a similar approach to learn lungs through an organ model, lung manikins were developed by Simbionix [64]. This model allows health workers to get fully trained on various lung zones. Using this model, they can simulate the condition of lungs in different clinical environments.

6.17   EMERGENCY DWELLINGS

During the COVID pandemic, hospitals were overloaded. Patients were not getting admission in hospitals and normally admitted in quarantine centers/home quarantine for initial treatment. This facility risks other asymptomatic patients and family members for COVID virus infection. Therefore, temporary isolation of each patient is required so that others won't get infected. This kind of temporary emergency dwellings/isolations can be made by 3D printing. This enables heath workers to isolate patients under quarantine, thus relaxing the overloaded medical infrastructures or hospitals. Compared to traditional construction methods, 3D printing of isolation units usually requires short times and lower costs. 3D-printed dwellings can also be easily transported and deployed to areas where there are poor medical facilities. A 3D-printed isolation ward was made by the Chinese company Winsun using a rapid 3D printing process [65]. This process has the capacity to manufacture 15 coronavirus isolation wards in a single day.

6.18 ROLE OF 2D MATERIALS IN 3D-PRINTED DEVICES

All the medical point-of-care devices printed by 3D printing are touched by patient, and their surfaces are prone to viral infection. Currently used polymer powder or filament for 3D printing can't inhibit transmission of virus. Therefore, the need for 2D materials, such as graphene, was used in filament during 3D printing. Normally, 2D materials like graphene are added in consumable filters of the mask or PPE, but there is a requirement of killing the virus on the component. This is normally done by UV light or autoclave. Flavio De Maio et al. [66] had added graphene in a PLA filament and printed point-of-care devices by FDM. It was demonstrated that the device can be sterilized by near-infrared light exposure. The viral particles on the surface of the component can be killed within 3 min of exposure. In another work related to the detection of virus by 3D-printed device [67], graphene was coated on 3D-printed electrodes. This was an advanced nanostructured bio-sensing platform that detects COVID-19 antibodies within seconds. The sensor is created by 3D printing of three-dimensional electrodes. These electrodes were coated with RGO along with specific viral antigens. COVID antibodies can be detected at a limit of 2.8×10^{-15}, and the data can be displayed on a smartphone.

In summary, 3D printing has lots of benefits in terms of custom-built design, complex structures, and minimum waste. A review of various 3D printing methods and the advantages and challenges were discussed. 3D printing has also contributed in the development of biomedical implants and devices according to patient-specific needs. The recent advancements of 3D printing in the biomedical field have contributed to building preventive, diagnostic, and treatment devices, also known as point-of-care devices, in the COVID pandemic. Collaboration of institutes and industries has enabled the prototyping and testing of new 3D-printed devices. The spread of coronavirus has enabled industries and governments to take quick decisions for 3D printing of critical tools and devices. As the 3D printing process is a rapid and low-cost manufacturing, 3D-printed point-of-care devices are user-friendly and are suitable for developing countries. In conclusion, 3D printing will play a very important role in manufacturing point-of-care devices for future pandemics also.

REFERENCES

1. Baylor University Libraries-3D Printing 101 [Internet]. [Cited 2022, May 12], Available from: http://libguides.baylor.edu/3dprinting101/process.
2. U.S. Patent 4,575,330. Apparatus for production of three-dimensional objects by stereolithography.
3. Berman B. 3-D printing: new industrial revolution. Business Horizons. 2012; 5(2):155–62.
4. Stansbury JW, Idacavage MJ. 3D printing with polymers: challenges among expanding options and opportunities. Dental Materials. 2016; 32(1):54–64.
5. Lee J-Y, An J, Chua CK. Fundamentals and applications of 3D printing for novel materials. Applied Materials Today. 2017; (7):120–33.
6. Araji Rad, Z, Prewett PD, Davies GJ. High-resolution two-photon polymerization: the most versatile technique for the fabrication of microneedle arrays. Microsystems & Nanoengineering. 2021; 7:71.
7. Qi G, Li Z, Wang Z, et al. Projection micro stereolithography based 3D printing and its applications. International Journal of Extreme Manufacturing. 2020; 2:022004.
8. Liu P, Guo Y, Wu Y, et al. A low-cost electrochemical metal 3D printer based on a microfluidic system for printing mesoscale objects. Crystals. 2020; 10:257.
9. Li Z, Rathore AS, Song C, et al. PrinTracker: fingerprinting 3D printers using commodity scanners. CCS'18: Proceedings of the 2018 ACM SIGSAC Conference on Computer and Communications Security, October 2018; 1306–23.
10. Mohamed OA, Masood SH, Bhowmik JL. Optimization of fused deposition modeling process parameters: a review of current research and future prospects. Advances in Manufacturing. 2015; 3(1):42–53.
11. Sood AK, Ohdar RK, Mahapatra SS. Parametric appraisal of mechanical property of fused deposition modelling processed parts. Materials & Design. 2010; 31(1):287–95.
12. Parandoush P, Lin D. A review on additive manufacturing of polymer-fiber composites. Composite Structures. 2017; 182:36–53.
13. Wang X, Jiang M, Zhou Z, et al. 3D printing of polymer matrix composites: a review and prospective. Composites Part B: Engineering. 2017; 110:442–58.
14. Utela B, Storti D, Anderson R, et al. A review of process development steps for new material systems in three dimensional printing (3DP). Journal of Manufacturing Processes. 2008; 10(2):96–104.
15. Wiberg, A. Towards design automation for additive manufacturing: a multidisciplinary optimization approach. Licentiate dissertation, Linkoping University Electronic Press, 2019.
16. Meiners W, Over C, Wissenbach K, et al. Direct generation of metal parts and tools by selective laser powder remelting (SLPR). Proceedings of the Solid Freeform Fabrication Symposium, University of Austin, 1999.
17. Lee H, Lim CHJ, Low MJ, et al. Lasers in additive manufacturing: a review. International Journal of Precision Engineering and Manufacturing-Green Technology. 2017; 4(3):307–22.
18. Kiran ASK, Veluru JB, Merum S, et al. Additive manufacturing technologies: an overview of challenges and perspective of using electrospraying. Nanocomposites. 2008; 4(4):190–214.
19. Dou R, Wang T, Guo Y, et al. Ink-jet printing of zirconia: coffee staining and line stability. Journal of the American Ceramic Society. 2011; 94(11):3787–92.
20. Travitzky N, Bonet A, Dermeik B, et al. Additive manufacturing of ceramic-based materials. Advanced Engineering Materials. 2014; 16(6):729–54.
21. Huang J, Qin Q, Wang J. A review of stereolithography: processes and systems. Processes. 2020; 8(9):1138.

22. Melchels FPW, Feijen J, Grijpma DW. A review on stereolithography and its applications in biomedical engineering. Biomaterials. 2010; 31(24):6121–30.

23. Manapat JZ, Chen Q, Ye P, et al. 3D printing of polymer nanocomposites via stereolithography. Macromolecular Materials and Engineering. 2017; 302(9):1600553.

24. Gibson I, Rosen D, Stucker B. Directed energy deposition processes. In: Additive Manufacturing Technologies: 3D Printing, Rapid Prototyping, and Direct Digital Manufacturing. New York, NY: Springer; 2015:245–68.

25. Lim JS, Oh WJ, Lee CM, et al. Selection of effective manufacturing conditions for directed energy deposition process using machine learning methods. Scientific Reports. 2021; 11:24169.

26. Ünal-Saewe T, Gahn L, Kittel J, et al. Process development for tip repair of complex shaped turbine blades with IN718. Procedia Manufacturing. 2020; 47:1050–7.

27. Wevolver. [Cited 2022, May 12], Available from: www.wevolver.com/article/laminated-object-manufacturing-creating-strength-with-layer.

28. Gibson I, Rosen D, Stucker B. Sheet lamination processes. In: Additive Manufacturing Technologies: 3D Printing, Rapid Prototyping, and Direct Digital Manufacturing. New York, NY: Springer; 2015:219–44.

29. J Li, T Monaghan, S Masurtschak, et al. Exploring the mechanical strength of additively manufactured metal structures with embedded electrical materials. Materials Science and Engineering: A. 2015; 639:474–81.

30. Li J, Monaghan T, Nguyen TT, et al. Multifunctional metal matrix composites with embedded printed electrical materials fabricated by ultrasonic additive manufacturing. Composites Part B: Engineering. 2017; 113:342–54.

31. Allen SM, Sachs EM. Three-dimensional printing of metal parts for tooling and other applications. Metals and Materials International. 2000; 6:589–94.

32. Mostafaei A, Elliott AM, Barnes JE, et al. Binder jet 3D printing—process parameters, materials, properties, modeling, and challenges. Progress in Materials Science. 2021; 119:100707.

33. Miyanaji H, Momenzadeh N, Yang L. Effect of printing speed on quality of printed parts in binder jetting process. Additive Manufacturing. 2018; 20:1–10.

34. Ziaee M, Tridas EM, Crane NB. Binder-jet printing of fine stainless steel powder with varied final density. JOM. 2017; 69:592–6.

35. Ventola CL. Medical applications for 3D printing: current and projected uses. Pharmacy and Therapeutics. 2014; 39(10):704.

36. Chen RK, Jin Y-A, Wensman J, et al. Additive manufacturing of custom orthoses and prostheses-a review. Additive Manufacturing. 2016; 12:77–89.

37. Jardini AL, Larosa MA, Maciel Filho R, et al. Cranial reconstruction: 3D biomodel and custom-built implant created using additive manufacturing. Journal of Cranio-Maxillofacial Surgery. 2014; 42(8):1877–84.

38. Banks J. Adding value in additive manufacturing: researchers in the United Kingdom and Europe look to 3D printing for customization. IEEE Pulse. 2013; 4(6):22–6.

39. Li J, Chen M, Fan X, et al. Recent advances in bioprinting techniques: approaches, applications and future prospects. Journal of Translational Medicine. 2016; 14(1):271.

40. Vanderburgh J, Sterling JA, Guelcher SA. 3D printing of tissue engineered constructs for in vitro modeling of disease progression and drug screening. Annals of Biomedical Engineering. 2017; 45(1):164–79.

41. Heinl P, Müller L, Körner C, Singer RF, Müller FA. Cellular Ti-6Al-4V structures with interconnected macro porosity for bone implants fabricated by selective electron beam melting. Acta Biomaterialia. 2008; 4(5):1536–44.

42. Khanoki SA, Pasini D. Multiscale design and multiobjective optimization of orthopedic hip implants with functionally graded cellular material. Journal of Biomechanical Engineering. 2012; 134(3):031004.

43. Erbel R, Di Mario C, Bartunek J, et al. Temporary scaffolding of coronary arteries with bioabsorbable magnesium stents: a prospective, non-randomized multi centre trial. The Lancet. 2007; 369(9576):1869–75.
44. Morrison RJ, Kashlan KN, Flanangan CL, et al. Regulatory considerations in the design and manufacturing of implantable 3D-printed medical devices. Clinical and Translational Science. 2015; 8(5):594–600.
45. Campbell I, Bourell D, Gibson I. Additive manufacturing: rapid prototyping comes of age. Rapid Prototyping Journal. 2012; 18(4):255–8.
46. Wang X, Xu S, Zhou S, et al. Topological design and additive manufacturing of porous metals for bone scaffolds and orthopaedic implants: a review. Biomaterials. 2016; 83:127–41.
47. Campbell T, Williams C, Ivanova O, et al. Could 3D printing change the world. In: Technologies, Potential, and Implications of Additive Manufacturing. Washington, DC: Atlantic Council; 2011.
48. Chen H, Guo J, Wang C, et al. Clinical characteristics and intrauterine vertical transmission potential of COVID-19 infection in nine pregnant women: a retrospective review of medical records. Lancet. 2020; 395(10226):809–15.
49. Rezaei M, RazaviBazaz S, Zhand S, et al. Point of care diagnostics in the age of COVID-19. Diagnostics. 2021; 11(1):9.
50. World Health Organization. Rational use of personal protective equipment for coronavirus disease (COVID-19). WHO. 2020.
51. He H, Gao M, Illés B, et al. 3D printed and electrospun transparent, hierarchical polylactic acid mask nanoporous filter. International Journal of Bioprinting. 2021; 6(4):309.
52. Iyengar K, Bahl S, Raju V, et al. Challenges and solutions in meeting up the urgent requirement of ventilators for COVID-19 patients. Diabetes & Metabolic Syndrome: Clinical Research & Reviews. 2020; 14:499–501.
53. Choong YYC, Tan HW, Patel D, et al. The global rise of 3D printing during the COVID-19 pandemic. Nature Reviews Materials. 2020; 5:637–9.
54. Radfar P, Bazaz SR, Mirakhorli F, et al. The role of 3D printing in the fight against COVID-19 outbreak. Journal of 3D Printing in Medicine. 2021; 5(1):51–60.
55. Isinnova SRL-Projects-Easy COVID [Internet]. [Cited 2022, May 12], Available from: www.isinnova.it/easy-covid19.
56. Formlabs. 3D printing to support COVID-19 response [Internet]. [Cited 2022, May 12], Available from: https://formlabs.com/covid-19-response.
57. COVID-19 call to action [Internet]. [Cited 2022, May 12], Available from: www.3dsystems. com/covid-19-response.
58. Stratasys Ltd. Stratasys helps: responding to the COVID-19 crisis [Internet]. [Cited 2022, May 12], Available from: www.stratasys.com/covid-19.
59. Mick P, Murphy R. Aerosol-generating otolaryngology procedures and the need for enhanced PPE during the COVID-19 pandemic: a literature review. Journal of Otolaryngology – Head & Neck Surgery. 2020; 49:1–10.
60. Amin D, Nguyen N, Roser SM, et al. 3D printing of face shields during COVID-19 pandemic: a technical note. Journal of Oral and Maxillofacial Surgery. 2020; 8:1275–8.
61. Temple J. How 3D printing could save lives in the coronavirus outbreak. MIT Technology Review. 2020. Available from: https://www.technologyreview.com/2020/03/27/950240/3d-printing-coronavirus-covid-19-medical-supplies-devices/#:~:text=Prisma%20Health%2C%20a%20major%20health,of%20the%20machines%20run%20tight.
62. Tooker A, Moya ML, Wang DN, et al. Performance of three-dimensional printed nasopharyngeal swabs for COVID-19 testing. MRS Bulletin. 2021; 46:813–21.
63. Michael Petch. 3D printing community responds to covid-19 and coronavirus resources [Internet]. [Cited 2022, May 12], Available from: https://3dprintingindustry.com/news/3d-printing-community-responds-to-covid-19-and-coronavirus-resources-169143.

64. Simbionix Simulators: COVID-19 MODULE [Internet]. [Cited 2022, May 12], Available from: https://simbionix.com/simulators/us-mentor/us-library-of-modules/u-s-covid-19.
65. Euronews Green [Internet]. [Cited 2022, May 12], Available from: www.euronews.com/green/2020/03/13/people-are-isolating-from-coronavirus-in-tiny-homes-built-in-two-hours.
66. De Maio F, Rosa E, Perini G, et al. 3D-printed graphene polylactic acid devices resistant to SARS-CoV-2: sunlight-mediated sterilization of additive manufactured objects. Carbon. 2022; 194:34–41.
67. Ali Md A, Hu C, Jahan S, et al. Sensing of COVID-19 antibodies in seconds via aerosol jet nanoprinted reduced-graphene-oxide-coated 3D electrodes. Advanced Materials. 2021; 33:2006647.

7 2D Materials–Modified Foams for COVID-19 and Other Healthcare Systems

Rajeev Kumar

Graphene is considered the thinnest, most promising 2D material due to its exceptional properties, like high carrier mobility of [Insert Equation Here]10,000 $cm^2V^{-1}S^{-1}$ at room temperature [1], large specific surface area of >2,500 m^2g^{-1} [2], good optical transparency of [Insert Equation Here]97.7%, high Young's modulus of [Insert Equation Here]1 TPa [3], and high thermal conductivity of 3,000–5,000 $Wm^{-1}K^{-1}$ [4]. This ultrathin graphene has been intensively explored in different fields, such as electronics [5] and optoelectronics [6], energy storage and generation [7], and environmental and biomedical applications [8]. In biomedical applications, graphene has been reported several times as a promising candidate for antimicrobial coating and biosensors due to its high electrocatalytic activity, excellent electrochemical and optical behavior, and high specific surface area. In graphene-based biosensors, researchers take graphene to enhance the electrochemical behavior of the sensors [9]. The high conductivity and the small bandgap of graphene promote electron conduction between graphene and the biomolecule. However, the poor dispersion of graphene and the inability to form a continuous conducting network make them an undesirable choice for sensor applications. The aforementioned problem can be solved by making a three-dimensional architecture of graphene, also called graphene foam. Graphene foams are lightweight, high-performance, porous materials containing an interconnected cell wall structure. It has attracted immense attention for its unique open porous structure and outstanding properties, such as ultralow density, enormous specific area, excellent electrical and thermal conductivity, good chemical and mechanical strength, along with corrosion-resistant properties [10, 11]. These extraordinary properties make graphene foam a promising candidate for various applications, such as in filters, catalyst supports, sorbents, scaffolds, EMI shielding, energy storage, antimicrobial and antiviral efficacy, electrochemical sensors, biosensors, and tissue engineering applications (Figure 7.1) [12]. The porosity and flexibility of graphene foam allow it to be used as a freestanding electrode, where a graphene network will help for charge transfer with small resistance. Further, the three-dimensional structure of graphene foam delivers a large specific surface area, which can prevent the agglomeration of graphene during the fabrication of graphene

DOI: 10.1201/9781003316381-7

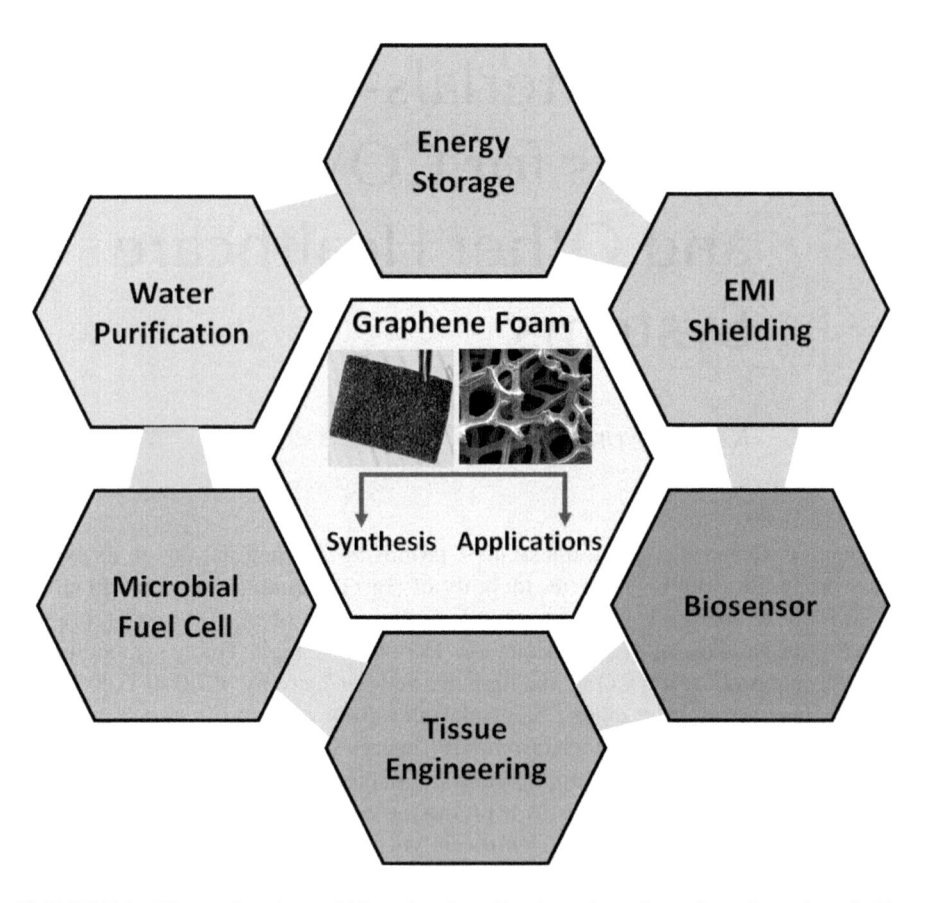

FIGURE 7.1 Illustrating the multifunctional application of graphene foam in various fields.

nanocomposites. Also, the 3D graphene foam has higher surface area than graphene; therefore, a large material surface will interact in enzyme and catalytic reactions. Coronaviruses (SARS-CoV-2) can enter wastewater through various pathways, such as wastewater discharged from hospitals and quarantine centers. Therefore, it is necessary to remove SARS-CoV-2 from wastewater. Due to the tunable pore size and physiochemical and photocatalytic nature of graphene foam, it is a potential material for SARS-CoV-2 filtration from wastewater. In this chapter, we will discuss the different methods of fabrication of graphene foam and their applications in biosensors, filters, and other healthcare systems.

7.1 SYNTHESIS OF GRAPHENE FOAM

7.1.1 TEMPLATE METHOD

In this method, metal foams, metal nanoparticles, metal oxide nanoparticles, and non-metal oxide particles are used as a template for the synthesis of graphene foam. The chemical vapor deposition (CVD) setup is used for the synthesis of graphene

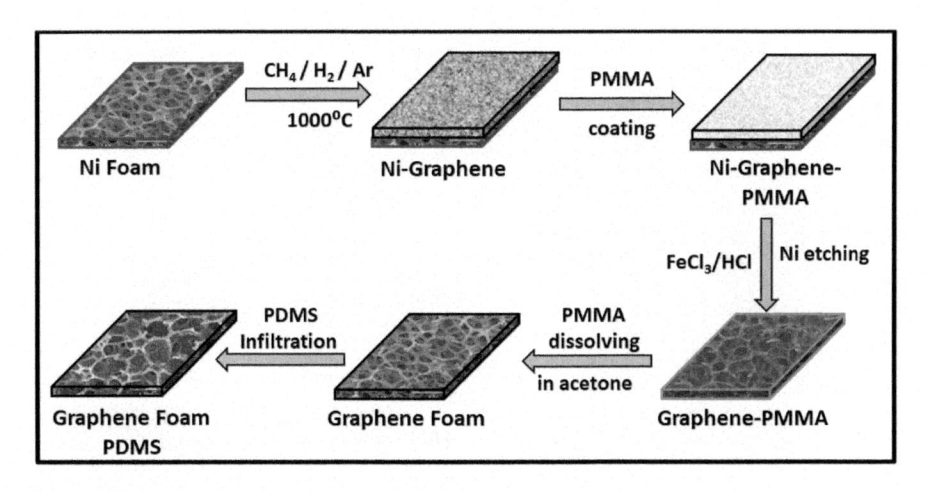

FIGURE 7.2   The schematic diagram for the construction of freestanding graphene foam [14].

foam, in which decomposition of carbon precursors, such as gaseous hydrocarbons (methane), liquid ethanol, and solid polymers, over the supporting template occurs [13]. In brief, metal foams, especially Ni and Cu, are generally used to synthesize graphene foam. The metal template is removed via chemical etching. Chen and his coworkers [14] used the CVD method to prepare flexible and very thin graphene foam on nickel foam. First, Ni foam with an interconnected porous structure was selected as a template for the growth of graphene foam. After that, the nickel foam was heated at 1000°C under Ar and H_2 and annealed for 5 min to remove the oxide layer. Then, CH_4 gas was pumped into the reaction tube at ambient pressure. At high temperatures, CH_4 was decomposed and Ni foam was covered with graphene sheets. Afterward, a thin layer of polymethyl methacrylate (PMMA) was deposited on graphene-coated Ni foam. Then, freestanding graphene/PMMA foam is obtained by etching the nickel foam in $FeCl_3$ and HCl solution. Finally, freestanding graphene foam was obtained by dissolving the PMMA in acetone. A freestanding graphene/PDMS composite was fabricated by the infiltration of PDMS on freestanding graphene foam, followed by curing at 80°C. The schematic diagram for the synthesis of graphene foam is shown in Figure 7.2.

The digital photographs of flexible graphene/PDMS composite foam are shown in Figures 7.3a–b, and its SEM images are shown in Figures 7.3c–d, which confirm that graphene composite foam is highly porous and similar in structure to nickel foam. Apart from methane sources, some other scientists have also used ethanol as a carbon source for graphene preparation; due to this, the emission of toxic gases can be avoided in the CVD process [15].

Further, Sun et al. reported that the graphene foam can also be obtained by etching Ni foam template without the coating of polymers [16]. Apart from the Ni foam template, the graphene foam can also be synthesized on non-metal porous structures, such as seashells [17]. Seashells are typically composed of $CaCO_3$. After annealing, the seashell is converted into a porous CaO, which provides a template for the synthesis of graphene foam. A macroporous structure of graphene can be obtained over

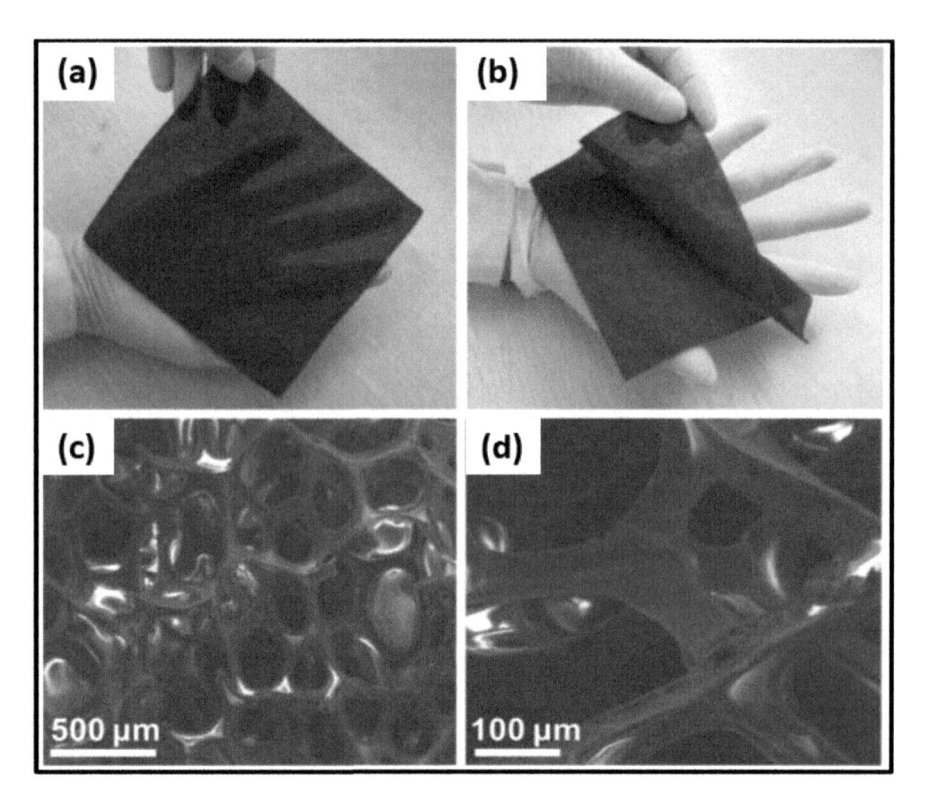

FIGURE 7.3 (a–b) Digital photographs and (c–d) SEM images of a flexible graphene foam developed through CVD.

Source: Reproduced with permission from ref. [14].

the porous CaO matrix using the CVD technique. After that, a freestanding 3D graphene foam structure is obtained by immersing the graphene–CaO matrix in an HCl solution. Samad et al. [18] used graphene oxide (GO) to produce graphene foam by a dip coating. Initially, the GO sheets are covered on a Ni foam surface by dip-coating under vacuum conditions. The GO-coated Ni foam is dried at 80°C and immersed in HI solution for the reduction of GO into rGO. Afterward, the rGO-coated Ni foam is put inside the 3 M hot HCl solution for 24 h for the removal of the nickel foam template, resulting in the synthesis of graphene foam (Figure 7.4). The prepared graphene foam is heated at 80°C for 2 h, and then graphene foam is impregnated with PDMS by vacuum infiltration and again heated at 80°C.

The graphene foam can also be produced by a polyurethane (PU) foam template. GO or rGO is used as a graphene precursor for the impregnation of PU foam. Shen et al. [19] developed ultra-low-density and compressible graphene-coated polymer foams using a PU urethane template. First, they synthesized the GO by modifying Hummer's method, and then 3 mg/mL of GO suspension are filled inside the PU foam. The GO-impregnated foam is then dried at 90°C to evaporate the water

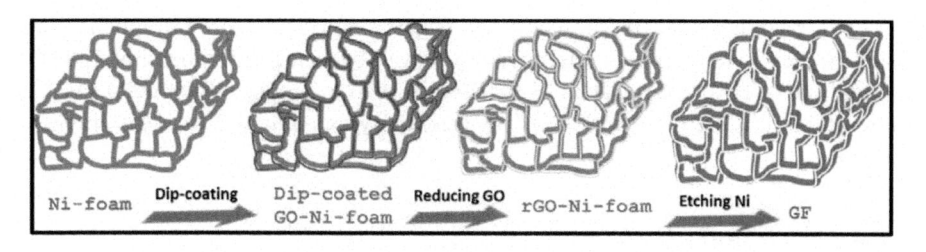

FIGURE 7.4   Complete fabrication process of graphene foam, which includes dip-coating of GO sheets onto the Ni foam and reducing of GO by thermal annealing.

Source: Reproduced with permission from ref. [18].

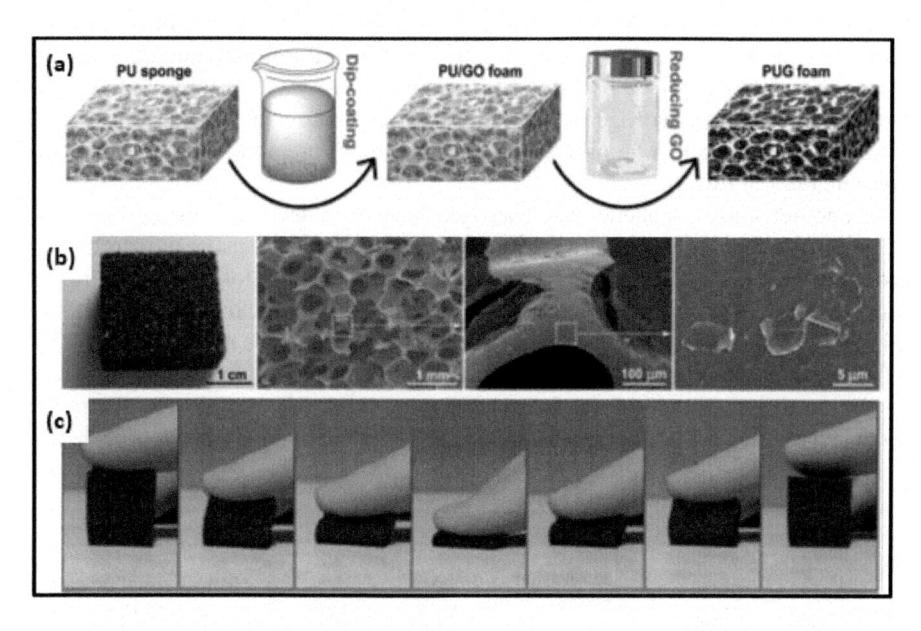

FIGURE 7.5   (a) Synthesis of the PUG foam, which includes impregnation of PU foam by GO sheets and then hydrothermal reduction by hydrazine hydrate; (b) optical and SEM images of PUG foam; and (c) compressing and releasing process of the PUG foam.

Source: Reproduced with permission from ref. [19].

completely. Finally, the dried foam is placed into a hydrazine monohydrate-filled Teflon vessel, which reduces the GO into rGO and obtains compressible polyure-thane graphene (PUG) foam, as shown in Figure 7.5.

Yadav et al. [20] used a PU foam template to prepare the graphite foam. In this investigation, a high-softening-point mesophase pitch is used as a carbon source, and PU foam as a template. The mesophase pitch was ground into a fine powder using ball milling and then mixed with water to form a homogeneous slurry. The slurry was impregnated into the PU foam slabs. The pitch-impregnated foams are then stabilized, carbonized, and finally graphitized at 2,400°C in an argon atmosphere to obtain graphite foam. In another work, Yuan et al. [21] prepared graphene foam using Nomex honeycomb as a template, followed by being carbonized at 1,000°C to obtain the CH structure. The CH structure is then impregnated with GO solution until all the cells are filled with GO. The freeze-drying method is used for graphene composite foam.

7.1.2   HYDROTHERMAL METHOD

The hydrothermal method is an effective technique for the synthesis of graphene foam, and this technique is also useful in fabricating the porous structure and decorating metal oxide nanoparticles and has good electrochemistry reactivity [22]. Deng et al. [23] prepared graphene foam from GO powder using the hydrothermal method. In this method, GO powder was well dispersed in DI water using ultrasonication. After mixing, the GO solution was placed into the Teflon-lined autoclave and put in an oven at 180°C for 12 h. After 12 h, the autoclave was left to cool down. The monolithic graphene hydrogel was freeze-dried and subsequently vacuum-dried to obtain graphene foam. The process of graphene foam preparation is schematically presented in Figure 7.6.

In another study, graphene-based aerogels have been developed using three types of carbohydrates, such as glucose, cyclodextrin, and chitosan, using a hydrothermal process [24]. Each carbohydrate was dissolved in acetic acid, followed by ultrasonication. After that, GO suspension was added to each mixture, and then the previous mixture was placed into a Teflon-lined stainless-steel autoclave and heated at 180°C. After cooling, the mixture was freeze-dried for the synthesis of graphene aerogel.

Taking the advantage of the freeze-drying method, Ma and his coworkers [25] produced MXene/reduced graphene oxide (MX/rGO) aerogel. In this approach, first,

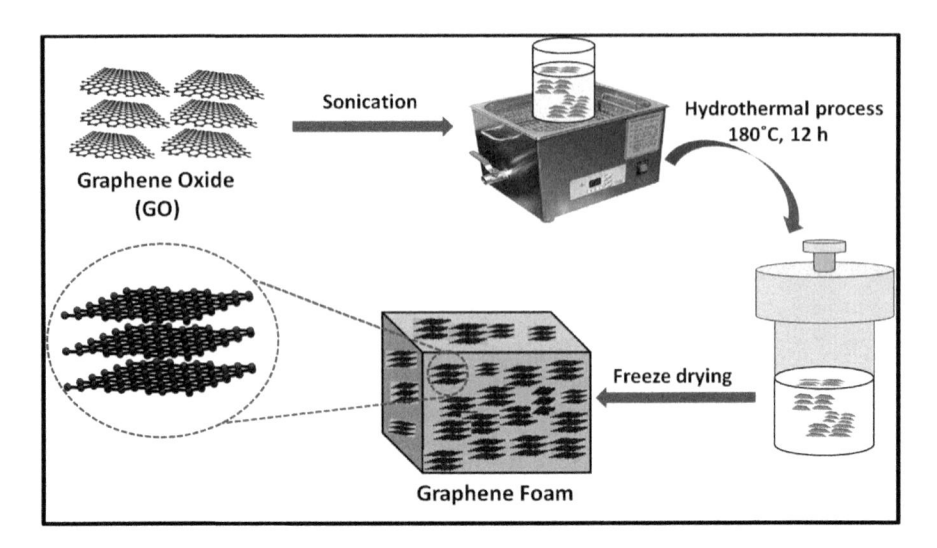

FIGURE 7.6   Schematic illustration for the fabrication of graphene foam using hydrothermal process.

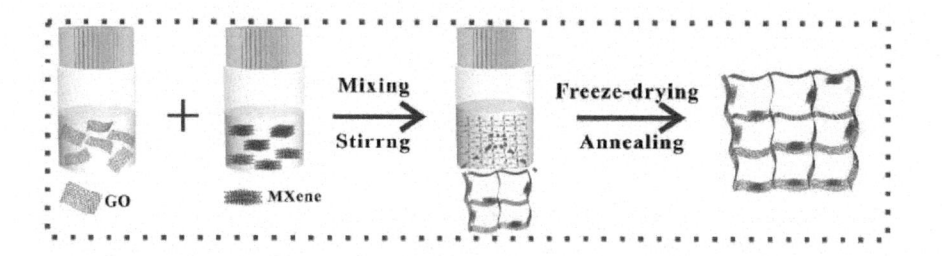

FIGURE 7.7 Synthesis of MXene/reduced graphene oxide aerogel using hydrothermal followed by freeze-drying method.

Source: Reproduced with permission from ref. [25].

a GO and MXene suspension was mixed and sonicated for 2 h. After that, the mixed solution was poured into a container and allowed for freeze-drying (−60°C, 1 Pa) to form the aerogel. Finally, the MXene/rGO aerogel was found by heating at 200°C under Ar/H$_2$, as shown in Figure 7.7.

Further, lightweight, robust, and porous MXene/polyimide aerogels with reversible compressibility were introduced by Dai et al. [26]. In this approach, they used poly amic acid (PAA) solution and MXene suspension with different volume ratios to form a uniform mixture. Then the mixture was poured into a silicone mold placed upon the copper bridge and freeze-dried to get the MXene/PAA aerogels. Subsequently, the prepared MXene/PAA aerogels were heated at 300°C under an argon atmosphere for complete imidization of PAA to PI. This results in an MXene/PI aerogel.

7.1.3 BLOWING AGENT METHOD

In this method, a blowing agent or foaming agent is generally used for the preparation of graphene foam. Foaming agents that produce gas via chemical reactions include ammonium bicarbonate, ammonium chloride, azodicarbonamide, titanium hydride, isocyanates, aluminum nitrate, etc. The mixture of graphene precursor and the foaming agent is heated up to the evaporation temperature of the foaming agents. Wang et al. [27] have utilized the sugar-blowing method to prepare 3D graphene architectures. In this method, a mixture of glucose with ammonium chloride was heated in a tubular furnace at 1,350°C for 3 h under an argon atmosphere. The glucose and ammonium chloride mixture was converted into a molten syrup and created many gases during polymerization. These produced gases form several large bubbles, which are further graphitized at high temperature and converted into graphene materials, as shown in Figure 7.8a. The SEM micrographs of graphene foam are shown in Figure 7.8b–d, which suggests that with increasing the heating rate, the cell size and strut width decrease.

Later on, Han et al. [28] developed a three-dimensional porous graphene (3D-PG) by thermal reduction of GO using the sugar-blowing method, as shown in Figure 7.9. In this study, NH4Cl and glucose were dissolved in DI water. Then, 5 ml of GO

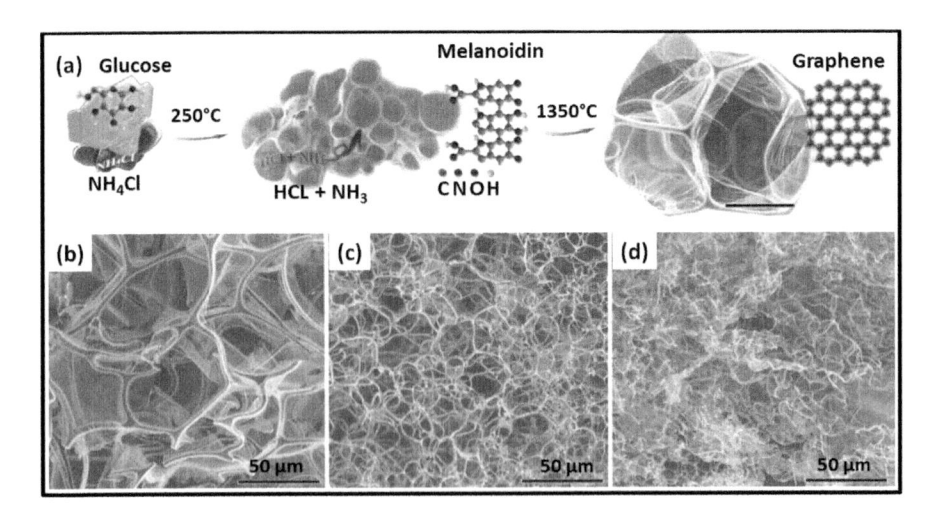

FIGURE 7.8 (a) Schematic figure for the preparation of graphene foam using the sugar-blowing method, and (b–d) SEM images of the bubble-shaped graphene foam prepared by the heating rates of 120, and 100°C/min, respectively.

Source: Reproduced with permission from ref. [27].

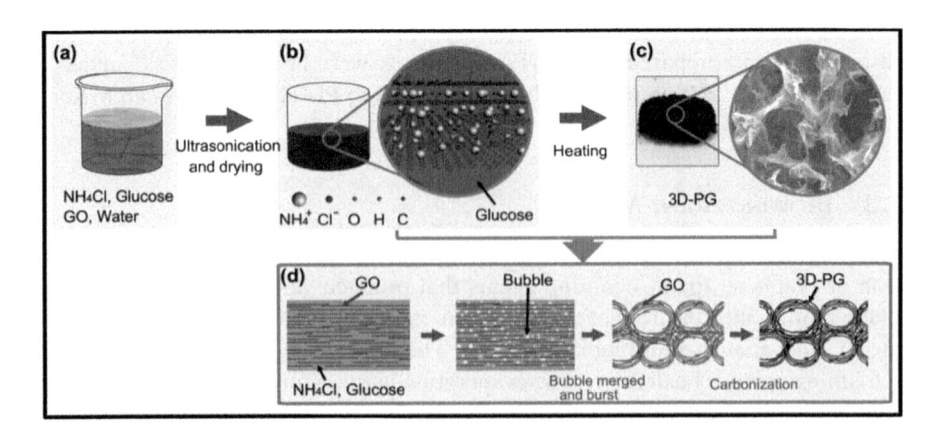

FIGURE 7.9 Schematic illustration for the synthesis of three-dimensional porous graphene (3D-PG): (a) mixing of GO, NH$_4$Cl, glucose, and water; (b) mixture of GO, NH$_4$Cl, and glucose; (c) SEM micrograph of 3D-PG; and (d) mechanism for the synthesis of 3D-PG.

Source: Reproduced with permission from ref. [28].

solution (2 mg/ml) were added to the solution and sonicated for 5 h. After that, the solution mixture was dried at 40°C. Then the mixture was transferred into a furnace and heated at 1,000°C for 2 h to produce the final three-dimensional porous graphene.

7.1.4. Powder Metallurgy Template Method

The powder metallurgy template method is the most common and inexpensive technique used for 3D foam synthesis. This method consists of many steps, such as mixing of carbon precursors and metal powder template, pressing, thermal decomposition, and etching of metal powder [29]. The synthesis of graphene foam using a powder metallurgy template has been reported by Sha et al. [30], as shown in Figure 7.10. In this method, Ni powder was used as a template, and sucrose was used as a carbon source. Both Ni and sugar powder were mixed in DI water and then dried at 120°C for the removal of water. Next, the Ni carbon powder was molded into pellets using hot-press. The pellets were put into a CDV furnace heated at 1,000°C under H_2/Ar atmosphere. After that, the graphene-coated pellets were cooled down to room temperature and then etched in an aqueous solution of $FeCl_3$ to remove the Ni, and then washed with DI water. Finally, freestanding graphene foam was obtained after drying in an oven.

In another method, Huang et al. [31] reported that spherical silica particles can also be used as a hard template to fabricate nanoporous graphene foams (NGFs) with controllable pore size (30–120 nm). In this study, silica sphere and GO were mixed in an aqueous solution; the hydrophobic nature of both GO and silica templates produces self-assembled lamellar-like structures (silica spheres wrapped between GO). The developed composite is then calcined to reduce the GO into graphene. After that, the silica sphere was removed by etching the composite in HF, and an NGF was obtained (Figure 7.11).

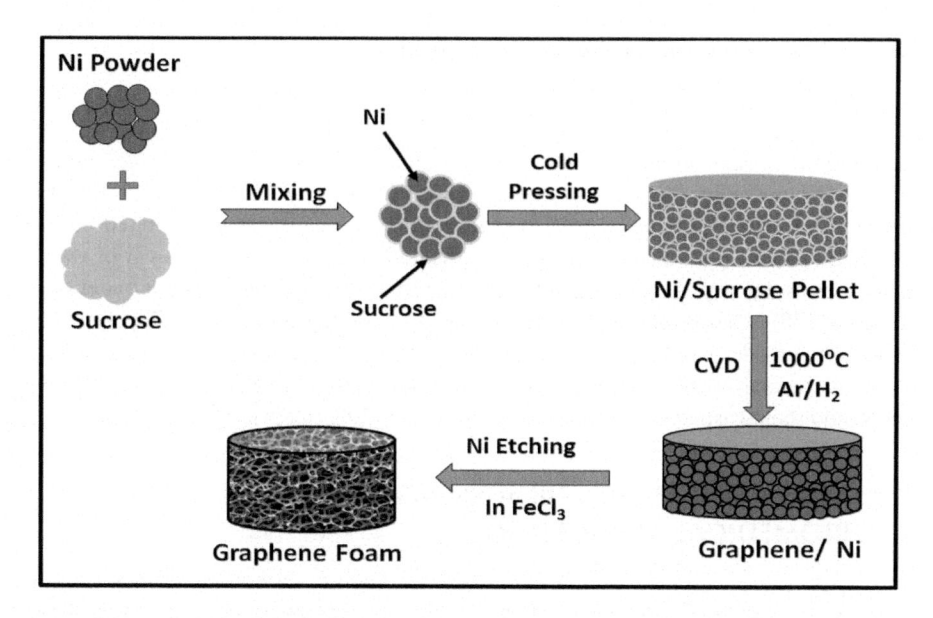

FIGURE 7.10   Synthesis of graphene foam using Ni as a powder metallurgy template.

Source: Reproduced with permission from ref. [30].

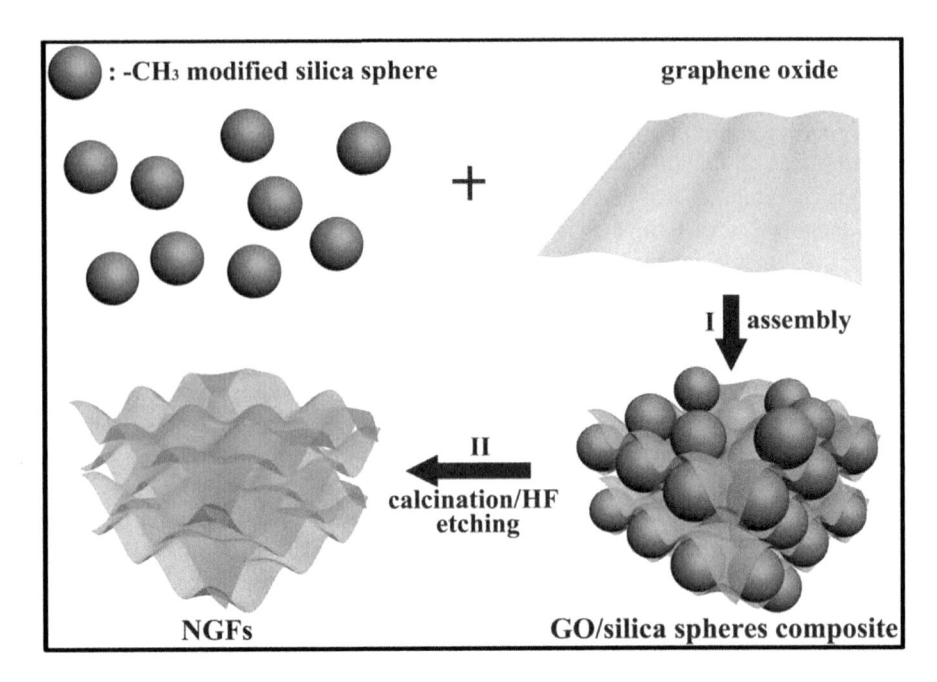

FIGURE 7.11 The schematic for the synthesis of nanoporous graphene foam (NGF).

Source: Reproduced with permission from ref. [31].

7.1.5 THREE-DIMENSIONAL PRINTING METHOD

Recently, 3D printing is a simple and effective method for the direct construction of a 3D bulk object. In 3D printing, first, 3D monoliths are designed using software, and then foam materials are heated and deposited layer by layer using a computer control printer. For example, Sha et al. [32] describe a 3D printing method to produce graphene foam. In this scheme, 3 g Ni powder and 0.5 g of sucrose were mixed in 200 mL of DI water, and then the mixture was heated at 80°C for the removal of water. After that, Ni sucrose powder was filled in a 1 mm thick steel case and put under pure H_2 gas (~175 sccm) pressure. A ZnSe window permitted an external CO_2 laser for Ni-sucrose powder irradiation inside the chamber, which converts the Ni-sucrose into Ni-graphene foam (GF). The 3D Ni-GF was immersed in 1 M $FeCl_3$ solution to remove the Ni and then washed with DI water to get freestanding 3D-printed GF (Figure 7.12).

7.2 GRAPHENE FOAMS FOR SENSORS AND BIOMEDICAL APPLICATIONS

In the past years, conventional carbon materials, such as carbon nanotubes, carbon nanofibers, graphene-reduced graphene oxide, and graphene quantum dots, have been purposed for sensing applications [33]. Nevertheless, these carbon materials are found in powder forms, which are usually coated onto another substrate to make

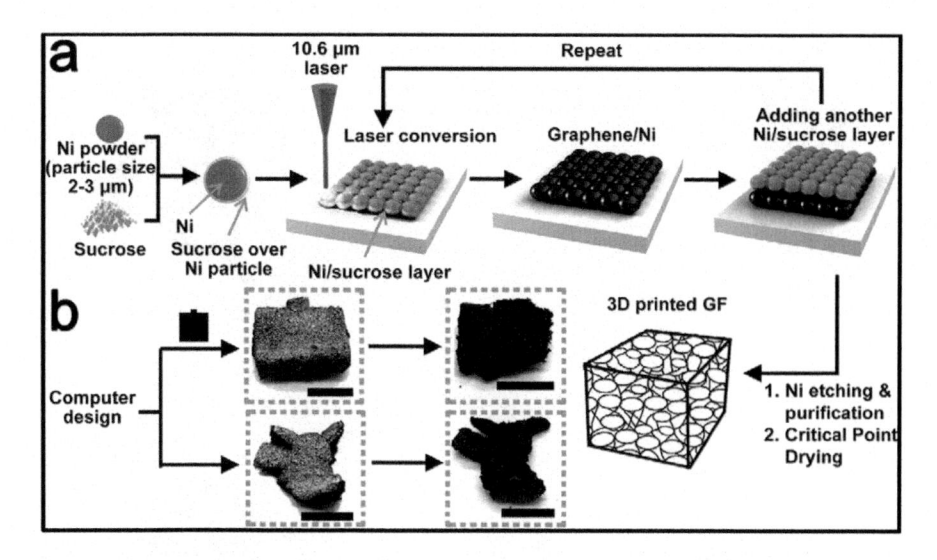

FIGURE 7.12 (a) Schematic illustration for the fabrication of 3D GF using 3D printing method, and (b) digital pictures of 3D-printed GF before and after removal of Ni.

Source: Reproduced with permission from ref. [32].

them conductive and mechanically robust. Also, a polymeric binder is required to hold the active material, which can increase the resistance in electrode and block the active sites. Therefore, the freestanding porous carbon is required, which can be directly used as a sensor. Recently, many efforts have been applied to graphene foam for sensors, due to its outstanding properties, such as high electrical and thermal conductivity, high mechanical strength, large surface area, and good biocompatibility [34]. Because of the large surface area, different active materials can be loaded into graphene foam. In addition, the high conductivity and good mechanical strength of graphene foam make it a conductive framework to support active materials and enhance the conductivity of sensing materials. Consequently, the open porous structure can generate more active sites and thus support fast charge transfer. So far, various electrochemical sensors with excellent performance, such as very low detection limit, small response time, and wide linear range, have been attained by using graphene foam and foam modified with different active materials.

Dong et al. [15] reported 3D graphene foam synthesis on Ni foam template using the CVD process. In their attempt, they observed that 3D graphene foam showed an outstanding sensitivity (619.6 μA m M^{-1} cm^{-2}) for dopamine, with a low detection limit (25 nM) and a linear response (up to ~25 μM). Similarly, 3D graphene foam (GF) with an interconnected network was fabricated by the CVD process using a Ni foam template. After that, GF was transferred into the indium tin oxide (ITO) glass by complete etching of the Ni foam template. The prepared GF/ITO electrode presented an excellent detection limit (3 nM) for uric acid [35]. The fabrication of the 3D GF/ITO electrode and oxidation-reduction reactions of UA and AA at the GF/ITO electrode is shown in Figure 7.13.

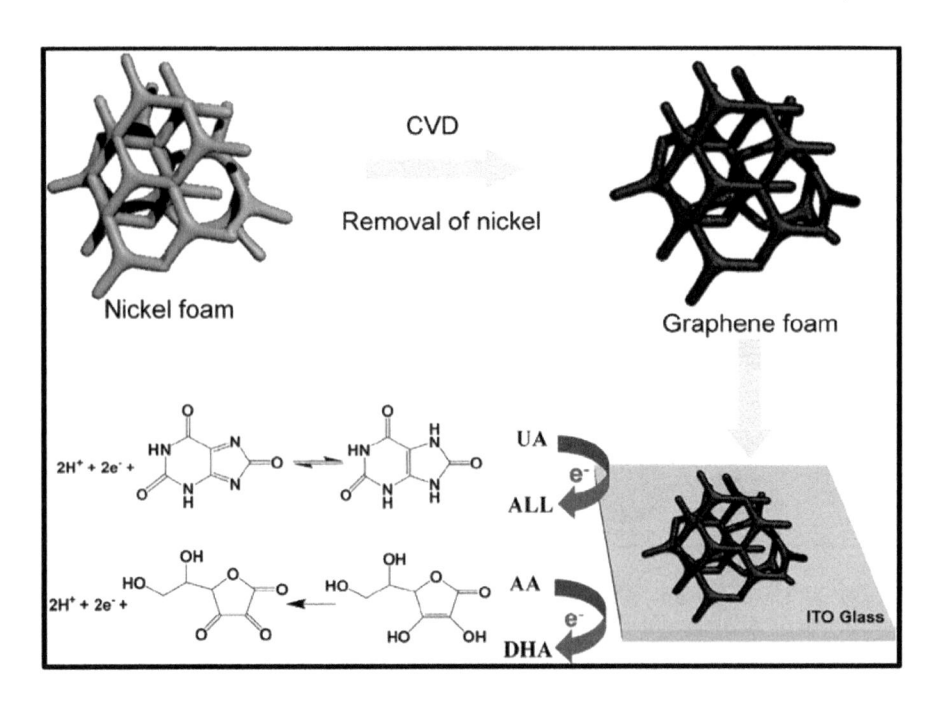

FIGURE 7.13 Schematic illustration for the fabrication of 3D GF/ITO electrode and oxidation-reduction reactions of UA and AA at GF/ITO electrode.

Source: Reproduced with permission from ref. [35]

Further, nitrogen-doped 3D foam (3D-NG) as an effective carrier of enzymes for biosensors has been investigated by Guo et al. [36] The 3D-NG was synthesized by the CVD method. The 3D-NG possesses high sensitivity for the sensing of glucose (226.24 µA mM^{-1} m^{-2}), which is two times higher than previously reported values. Similarly, Wang et al. [37] used zeolite Ni-MCM-22 as the template to prepare biocompatible monocellular graphene foam (MGF) for the immobilization and biosensing of glucose oxidase (GOD). Through the experiment, it is found that pristine MGF exhibited a fast electron transfer with a rate constant of 4.8 s^{-1}. It is also observed that the contribution of MGF in terms of detection and sensitivity are 0.25 mM and 2.87 µA mM^{-1} cm^{-2}, respectively.

To further improve the sensitivity and achieve the lowest detection limit in graphene foam–based sensors, different transition metal nanoparticles, such as silver, gold, and platinum, were used [38]. As metal nanoparticle surfaces have a large surface-to-volume ratio and binding sites, this results in fast communication between an enzyme and a nanoparticle for signal transduction in biosensing. Therefore, platinum–ruthenium bimetallic nanocatalysts–supported graphene foam (PtRu/3DGF) have been utilized for hydrogen peroxide–based electrochemical biosensors [39]. First, they fabricated the graphene foam on Ni template via CVD method, and then the prepared graphene foam was modified with PtRu nanoparticle catalyst using the borohydride reduction method. From Figure 7.14a, it can be seen that graphene foam

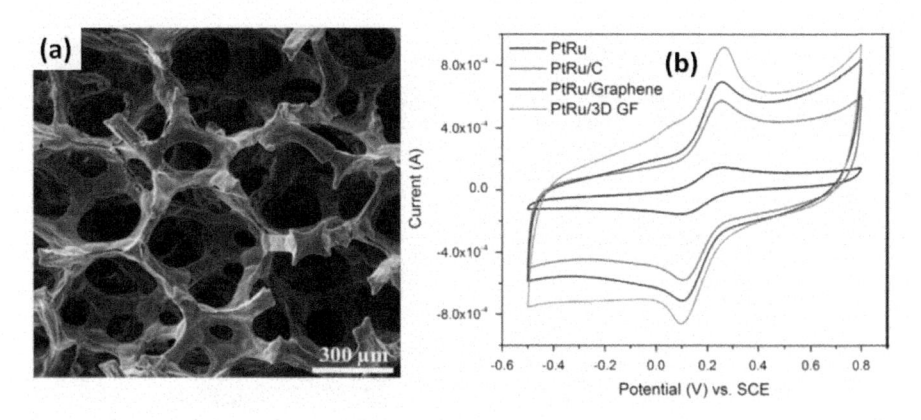

FIGURE 7.14   (a) SEM image of graphene foam and (b) cyclic voltammetry of PtRu nanocatalysts with different carbon materials in potassium ferricyanide and potassium ferrocyanide.

Source: Reproduced with permission from ref. [39].

exhibits a macroporous structure with a pore size of 50–250 μm. The cyclic voltammetry (CV) of PtRu nanocatalyst with different carbon materials was performed in potassium ferricyanide and potassium ferrocyanide, as shown in Figure 7.14b. From this figure, it has been observed that the redox peak intensity of PtRu nanocatalyst was low, and the peak-to-peak potential separation ($\Delta Ep = Epa - Epc$) was 172 mV. On the other hand, the PtRu/C, PtRu/graphene, and PtRu/3DGF nanocatalysts exhibited low peak separation, that is, 152, 155, and 166 mV, respectively. In addition, the PtRu/3DGF presented a high anodic and cathodic peak current and excellent electron and mass transfer rate due to the 3D structure and high surface area of graphene foam. Therefore, PtRu/3DGF nanocatalyst showed a high sensitivity (1,023.1 mA mM^{-1} cm^{-2}) and a low detection limit (0.04 mM) for H_2O_2.

In addition, some interesting research has been carried out on graphene foam combined with other nanomaterials, such as CNTs and metal nanoparticles [40]. Due to the synergistic effects of 3D graphene and the nanomaterials, these composites could be the best choice for enhancing electrochemical-sensing performance. Huang et al. [41] used freestanding 3D graphene foam (GF) with carbon nanotubes (CNTs) and gold nanoparticles (GNPs) electrodes for the determination of dopamine (DA) in brain tissue of SD rats and uric acid (UA) in human urine. The GF network and CNTs not only provide a high surface area but also offer an excellent conductive network to facilitate electron transfer in GNPs. The freestanding GF/CNTs/GNPs electrodes exhibit remarkable sensitivity of 12.72 and 3.36 μA μM^{-1} cm^{-2} and low detection limits of 1.36 and 33.03 nM for DA and UA, respectively. In another study, Yang et al. [42] fabricated 3D graphene foam covered with Ni particles (Gr-NiP) using CVD and stamp-transfer processes for stretchable sensor applications. The prepared Gr-NiP has 80% stretchability and also showed a low detection limit (<1%) and excellent linearity ($R^2 = 0.997$).

Apart from metal nanoparticles, some metal oxides can also be introduced into the graphene foam to form a synergistic effect between graphene foam and nanoparticles, which can improve the electron transfer rate in sensors [43]. In this series,

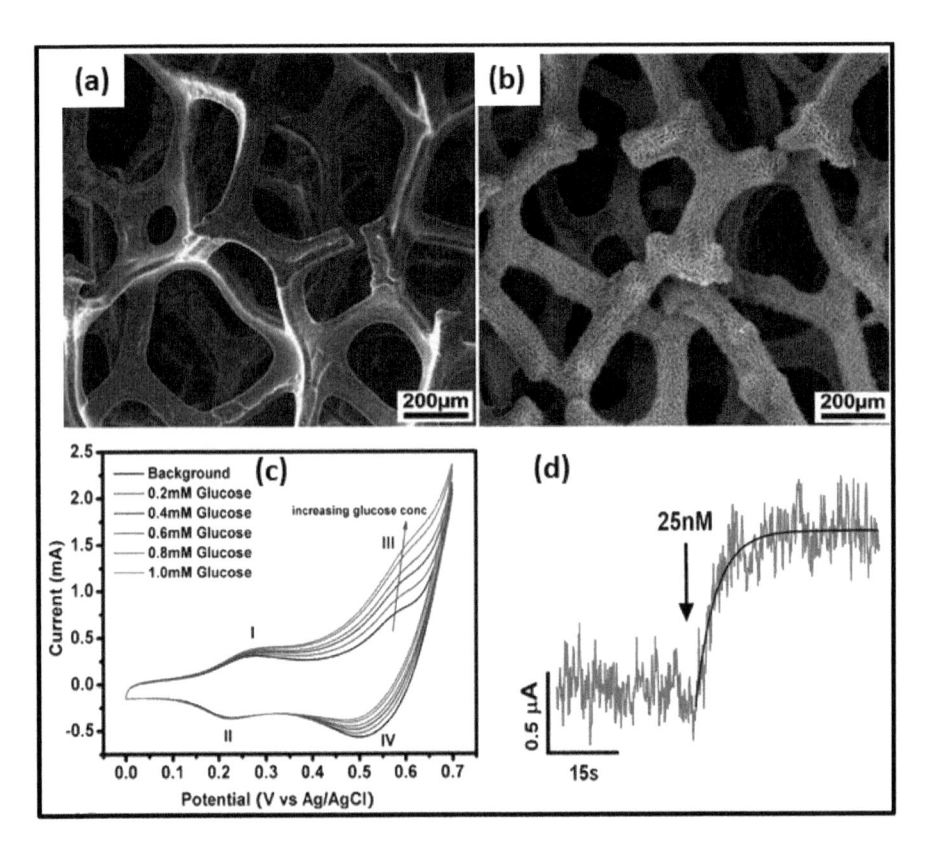

FIGURE 7.15 SEM images of (a) graphene foam and (b) 3D graphene/Co_3O_4 composite foam, (c) CV curves at the scan rate of 20 for different concentrations (0, 0.2, 0.4, 0.6, 0.8, and 1 mM) of glucose, and (d) amperometric response of graphene/Co_3O_4 nanowire composite for glucose.

Source: Reproduced with permission from ref. [44].

3D graphene/Co_3O_4 composite foam has been reported for enzymeless glucose detection [44]. In this study, first, graphene foam was prepared on a Ni foam template via the CVD method, and then prepared graphene foam was modified with Co_3O_4 nanowires using the hydrothermal method. Figure 7.15a shows the SEM image of graphene foam which exhibited a 3D porous structure with a pore size of 100–120 μm. After the growth of Co_3O_4, the graphene foam is fully and uniformly decorated with Co_3O_4 nanowires, as shown in Figure 7.15b. CV curves of 3D graphene/Co_3O_4 composites in 0.1 M NaOH and different concentrations of glucose, at the scan rate of 20 mV/s, are presented in Figure 7.15c. From this figure, it was found that the oxidation current at peak III (at ~0.58 V) was enhanced by the addition of glucose in NaOH; however, the current at the peak I (at ~0.25 V) remains constant. This may be due to glucose oxidation to gluconolactone catalyzed by conversion of CoO_2 to CoOOH (redox pair III/IV), which can be written as follows:

$$2CoO_2 + C_6H_{12}O_6(glucose) \rightarrow 2CoOOH + C_6H_{10}O_6 \qquad (1)$$

As a result, the 3D graphene/Co_3O_4 composites electrode showed a low detection limit of 25 nM and a wide linear range for glucose sensing (Figure 7.15d).

In another report, Si et al. [45] demonstrated hierarchically structured Mn_3O_4 grown on 3D graphene foam (3DGF) for non-enzymatic determination of glucose and H_2O_2, and also useful for health care and the food industry. They observed that Mn_3O_4/3DGF-based glucose sensor had a large linear detection range of 0.1–8 mM and also showed excellent sensitivity for the detection of H_2O_2. The sensing performance of Mn_3O_4/3DGF is mainly due to the electrocatalytic properties of Mn_3O_4, the conductivity of graphene, and the high specific surface area of the composite. 3D graphene foam (GF) decorated with vertically aligned ZnO nanowire arrays (ZnO NWAs) was used for the detection of uric acid (UA), dopamine (DA), and ascorbic acid (AA) by a differential pulse voltammetry method [46]. The prepared ZnO NWA/GF electrode delivered a high surface area and low detection limit of 1 nM for UA and DA. The same material is also used to determine the UA levels in the serum of patients with Parkinson's disease (PD). It was confirmed that a patient with Parkinson's disease (PD) has a 25% lower UA level than a healthy person. Further, Yue et al. [47] have discussed the significance of graphene foam with covered ZnO nanosheet spheres (ZnO NSSs) for uric acid detection. The ZnO NSSs/GF electrode demonstrates better sensitivity of 0.66 $\mu A\ \mu M^{-1}$ and a lower detection limit of 1 μM for uric acid present in human urine samples.

Again, in this series, Au nanoparticles-ZnO nanocone arrays/graphene foam (Au-ZnO NCAs/GF) electrode was prepared by Yue et al. [48]. In this experiment, first, they prepared graphene foam via CVD, and then ZnO nanoparticle arrays were synthesized on its surface by the hydrothermal process, followed by the self-assembling of Au nanoparticles with carboxyl groups, as shown in Figure 7.16. It was observed

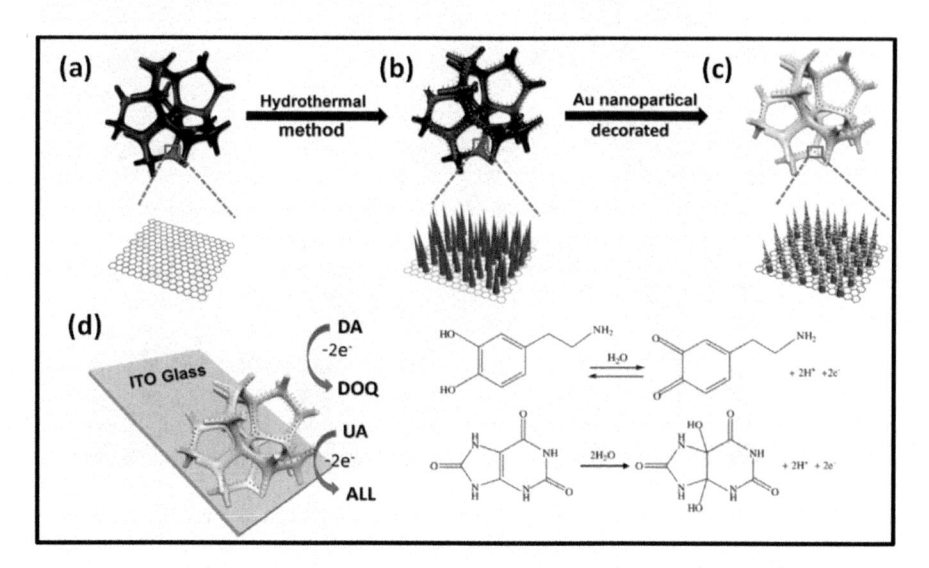

FIGURE 7.16 (a) GF, (b) ZnO NCAs on GF, (c) Au nanoparticles–decorated ZnO NCAs/GF, and (d) redox reactions of DA and UA at the Au-ZnO NCAs/GF electrode.

Source: Reproduced with permission from ref. [48].

that the electrode showed a very high sensitivity (4.36 µA µM⁻¹) and a low detection limit (0.04 µM) for dopamine.

Most recently, graphene foam/hematite (GF/α-FeO) nanowire arrays heterostructured nanocomposite (HNC) electrode was used for the detection of glucose [49]. The obtained HCN electrode has a high surface area, which favors the immobilization of large amounts of glucose oxidase. The detection limit of 71.6 µM and sensitivity of 20.03 µA mM⁻¹ cm⁻² are achieved in the HCN electrode.

Besides the sensors, graphene foam provides a biocompatible scaffold for cell therapy and tissue engineering. As we already discussed, graphene foam has a highly porous and interconnected structure, which significantly increases the surface area for cell growth. In this direction, Krueger et al. [50] investigate the culture of C2C12 murine myoblast cells on 3D graphene foam prepared by CVD and used as a scaffold for the growth of muscle cells. The growth of C2C12 cells into functional myotubes on both bare and laminin-decorated graphene foam has been confirmed by immunostaining and confocal microscopy. It was observed that both bare and laminin-decorated graphene foams provide a biocompatible scaffold for the growth and differentiation of muscle cells, as shown in Figure 7.17.

Also, the graphene foam polydimethylsiloxane (GF-PDMS) composite by Samad et al. [18] confirmed its ability to sense both compressive and bending strains in the form of change in electrical resistance. The prepared GF-PDMS composites (Figure 7.18a) exhibited different sensitivity to bending and compression. Figure 7.18b

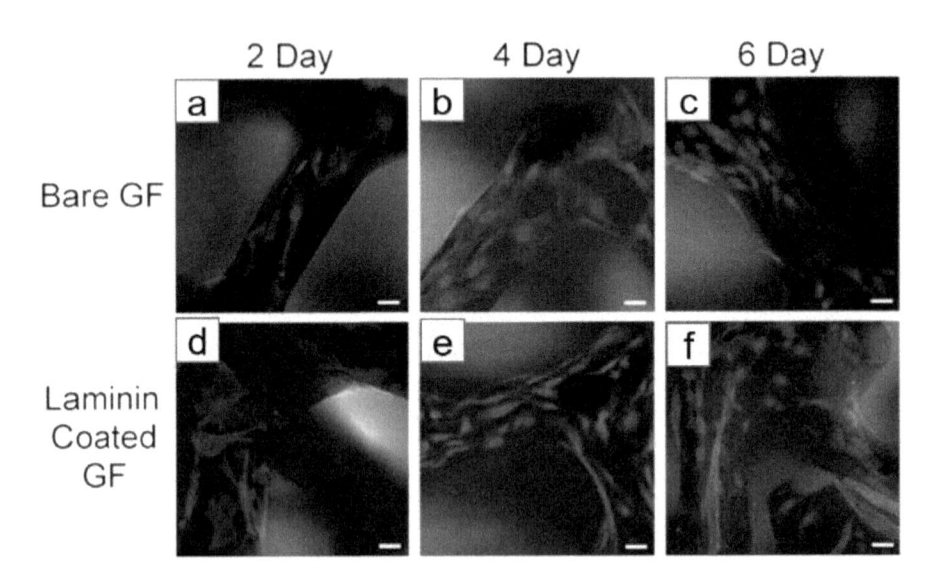

FIGURE 7.17 Graphene foam and laminin-coated graphene foam as a biocompatible scaffold for the growth and differentiation of muscle cells for different periods. The elongated blue nuclei represent the ongoing differentiation. Blue represents the nuclei of Hoechst. Red represents the actin of Alexa Fluor 546 phalloidin.

Source: Reproduced with permission from ref. [50].

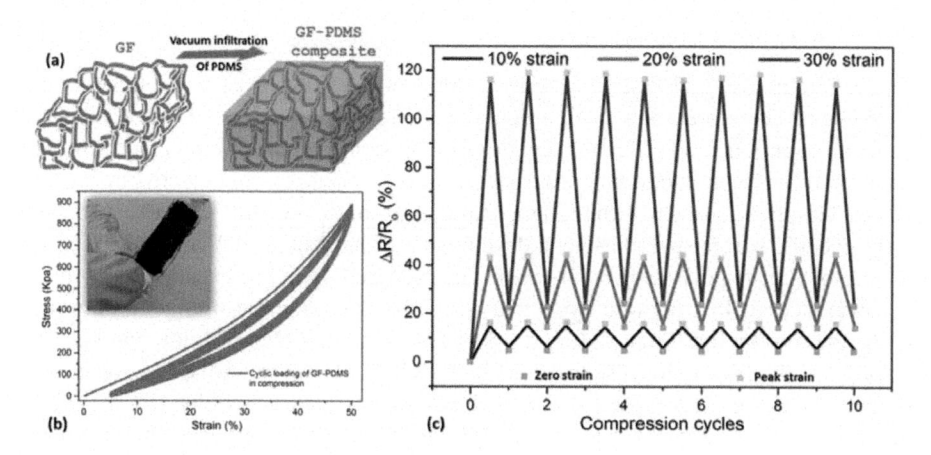

FIGURE 7.18   (a) Schematic figure for the impregnation of PDMS inside the GF, (b) stress-strain curve of a GF-PDMS composite with inset image of GF-PDMS composite, (c) change in resistance of the GF-PDMS composite as a function of compressive cycles.

Source: Reproduced with permission from ref. [18].

shows the mechanical integrity (loading–unloading compression cycle nos. 1–500) of GF-PDMS composites, which confirms that the hysteresis is almost the same and no significant internal structure has been changed. When applying a 30% compressive strain on the GF-PDMS composite, the resistance of the composite increased to ~120% of its original value, as shown in Figure 7.18c.

7.3   GRAPHENE FOAM FOR FILTRATION

The rapid development which includes industrialization and urbanization has generated various toxic organic compounds and heavy metal ions in surface water globally. These toxic impurities can cause several harmful effects on human health and animals [51]. Therefore, the removal of these toxic impurities is mandatory for the purification of water. Adsorption is the most widely used, economic, and versatile technique for water purification, which allows efficient removal of various toxic impurities from water. In past decades, different adsorbent with high efficiency and good recyclability have been introduced [52, 53]. For adsorption, graphene foam has become a popular alternative for water purification because of its excellent surface area and large interconnected open pores that provide sufficient transportation channels for the diffusion of pollutants, resulting in fast kinetics [54–56]. In addition, the hydrophobic nature and the π conjugate system offer a specific interaction with organic pollutants. Due to excellent electrical and thermal conductivity and good chemical resistance, the graphene foam can be recycled and regenerated by heating and electrochemical oxidation [57]. Thus, the graphene foam can be used as a superadsorbent in environmental remediation to remove toxic metals and organic pollutants [56, 58].

7.3.1 Removing Organic Dyes

The textile industry discharges about 60,000 t of dyes into wastewaters every year worldwide. Therefore, there is an urgent need to remove these dyes from wastewater using an economic and efficient way. In this direction, graphene foam can be used as an effective sorbent due to its porous structure, π–π stacking, and electrostatic interaction. Graphene aerogel with high mechanical strength, uniform pore structure, and large specific surface area has been developed by Zhang et al. [59] and is applied for the removal of organic dyes from water. The prepared foam has excellent adsorption performance against organic dyes, that is, methyl blue, methyl orange, and orange G. It can be seen that after adding 2 mg of GA in methyl blue solution and keeping it for 12 h, the blue color disappears (Figure 7.19a). However, in the case of methyl orange, the adsorption capacity of the GA is weaker than methyl blue (Figure 7.19b) but stronger than orange G (Figure 7.19c). The high absorption capacity of GA for methyl blue is mainly due to the highly porous structure and large surface area of GA, which leads to an increase in the contact angle of the dye molecule.

Mei et al. [60] used graphene aerogel modified with ZnO (GA-ZnO) for the efficient removal of methylene blue (MB) dye from wastewater. It was found that GA-ZnO has excellent adsorption performance (94.2%) for MB, and total adsorption capacity has also been increased up to 97.6% under visible light irradiation. Similarly, graphene oxide aerogel (GA) functionalized with polydopamine (PDA) and polyethyleneimine (PEI) was prepared and used for the adsorption of methyl orange and amaranth [61]. The prepared aerogel exhibits adsorption capacity of 202.8 and 196.7 mg/g for methyl orange and amaranth, respectively. Very recently, Liu et al. prepared carboxymethyl cellulose/graphene aerogel composite beads (CMC/GA) for the removal of methylene blue (MB) from contaminated water [62]. It was found that the removal rate of MB from the water was 90% even after 30 times adsorption regeneration cycles.

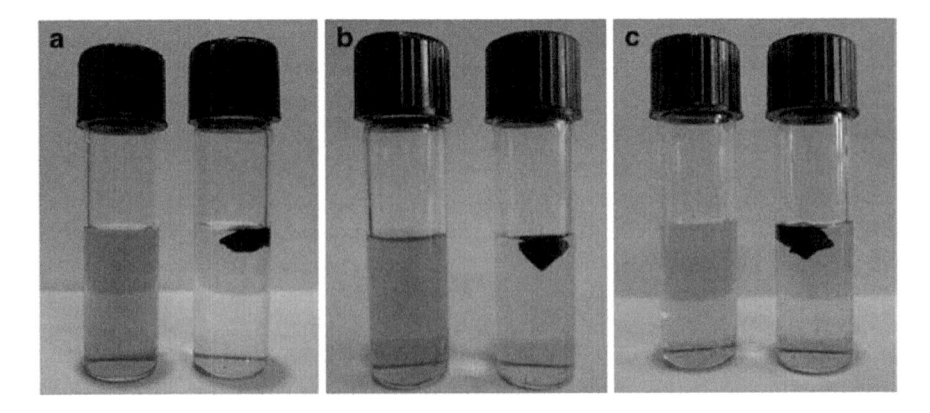

FIGURE 7.19 Photographs of the removal of dyes from water: (a) methyl blue, (b) methyl orange, and (c) orange G before (left) and after (right).

Source: Reproduced with permission from ref. [59].

7.3.2 Oil–Water Separation

At the same time, the hydrophobic nature and abundant negatively charged oxygen-containing groups in graphene foam make them a potential candidate for adsorption and separation of crude oil from water [63]. In addition, the functional group present on the surface of graphene foam is helpful for the filtration of water, gasoline, kerosene, olive oil, etc. [64]. For instance, Sun et al. [65] prepared ultra-flyweight aerogels (UFA) (Figure 7.20a) with graphene oxide and CNTs using the freeze-drying method. Figures 7.20b and c show the interconnected, porous graphene structure, which was covered the entangled CNTs network. The prepared UFA has shown an ultrahigh absorption capacity of 215–913 times its weight when it was immersed in a toluene and water mixture for 5 s, as shown in Figure 7.20c.

In another study, Yang et al. [66] demonstrate that graphene foam prepared through polystyrene particles as a sacrificial template has hierarchically porous structures, which help separate the oil from the oil–water mixture. Further, the sorption capacity of reduced-graphene oxide foam has been improved by introducing Fe_3O_4 magnetic nanoparticle (MNP) [22]. Motor oil was used to determine the adsorption capacity of the rGO-MNP hybrid foams. It was found that the rGO-MNP hybrid foams have low sorption capacity than the pristine rGO foam. There are two reasons for the decrease in the sorption capacity of rGO-MNP hybrid foams. First, increase the hydrophilicity of rGO-MNP hybrid foams due to the existence of oxygen molecules in MNP, and second, MNP slightly decreased the surface area and the pore volume of rGO-MNP hybrid foams.

Recently, Krebsz et al. [67] have synthesized bio-graphene foam (bGF) from renewable sources (glucose, citric acid, urea, and GO) using carbonization at 900°C under an argon atmosphere. The prepared bGF has shown a uniform porous structure and large specific surface area (805 m^2 g^{-1}), which is highly desirable for the

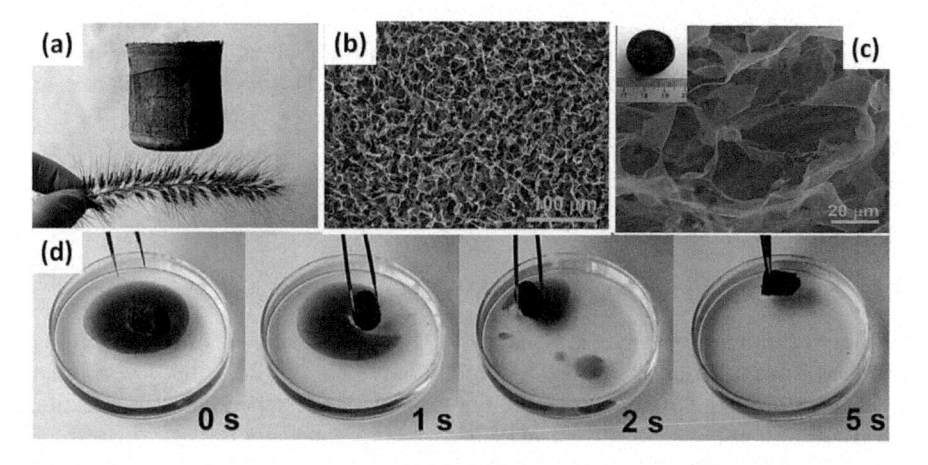

FIGURE 7.20 (a) As prepared ultra-flyweight aerogel (UFA), (b–c) SEM images of UFAs, and (d) sorption of toluene on the water by the UFA within 5 s.

Source: Reproduced with permission from ref. [65].

adsorption of oil from contaminated water. The adsorption capacity of bGF was measured to be 99.1% and 98.8% for toluene and petroleum, respectively.

7.3.3 Removing Heavy Metal Ions

Heavy metal ions are also discharged from industrial and agricultural fields. These metal ions are very toxic and harmful to the environment, as well as the health of human beings. To remove these metal ions from water, various cost-effective methods have been investigated. For example, GO or rGO aerogel surfaces have negatively charged oxygen-containing groups, which are responsible for the adsorption of heavy metal ions. Therefore, GO and rGO modified with magnetic materials has been employed to remove the As^{3+}, Cr^{6+}, Co^{2+}, Cu^{2+}, and U^{6+} from contaminated water [57]. GO aerogel has shown porous architecture and good adsorption capacity toward both positively charged Cu^{2+} (19 mg g^{-1}) [68] and negatively charged chromate $Cr_2O_7^{2-}$ (140 mg g^{-1}) [69]. In research published by Kabiri et al. [70], graphene-diatom silica aerogels were developed for mercury adsorption. In this study, the adsorption capacity of aerogel was investigated as a function of mercury concentration. The Langmuir model was applied to calculate the adsorption capacity of aerogel, which was found to be >500 mg of mercury/g of adsorbent. Similarly, another example of polyethyleneimine (PEI) cross-linked GO aerogel was used for the adsorption of As(V) and As(III) [71]. The prepared aerogel exhibits adsorption capacities of 4.80 and 4.26 mg g^{-1} for As(V) and As(III), respectively. Further, Gao et al. [72] reported the research on reduced graphene oxide (rGO) aerogels for the successful removal of Pb(II) from water. In addition, to find stability and recyclability, rGO aerogel has also been applied for adsorption-desorption cycle experiments. Recently, Zhu et al. [73] used a one-step method to fabricate the tannin-reduced graphene aerogel (TRGA). The prepared TRGA has shown low density, high porosity, and high specific surface area (122.2 m^2/g). The maximum adsorption capacity of TRGA reached 803.84 mg g^{-1} at pH 4 for Pb(II) and 395.80 mg g^{-1} at pH 4 for Cd(II). Apart from heavy metals, graphene foams are also suitable for the removal of radioactive materials from water. In this direction, various graphene-based aerogels have been investigated for the removal of U(VI), Sr^{2+}, Cs^+, etc. [74–76]. Li et al. [77] synthesized rGO aerogel from graphene oxide sheets using fungus hypha as a template. Due to its high surface area, the synthesized aerogel has shown an excellent adsorption capacity of 288.42 mg/g for uranium VI (U VI).

7.3.4 COVID-19 Filtration from Wastewater

In March 2020, coronavirus disease 2019 (COVID-19) has been declared a pandemic and has affected the worldwide community [78, 79]. Recently, several countries, like the United States, Australia, France, Italy, and Netherland, have identified the transmission of coronavirus (SARS-CoV-2) in wastewater [80, 81]. In addition, the SARS-CoV-2 was confirmed in the urine samples of infected patients [82]. Similarly, the SARS-CoV-2 was also reported in the hospital sewage and community wastewater [83, 84]. The SARS-CoV-2 can enter the water systems through various pathways, such as wastewater discharged from hospital and quarantine centers,

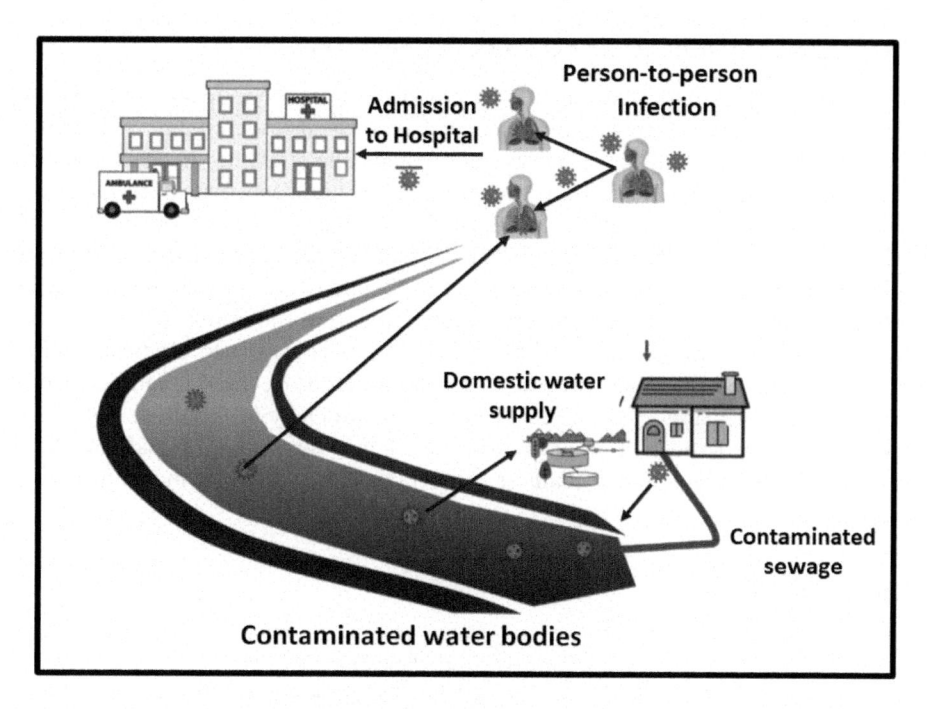

FIGURE 7.21   Sources and paths of SARS-CoV-2 in water systems.

Source: Reproduced with permission from ref. [85].

and can produce several possible transmissions of SARS-CoV-2 in human beings (Figure 7.21) [85].

Thus, wastewater treatment and the removal of SARS-CoV-2 from water are very important. It was also mentioned that the size of SARS-CoV-2 is ~140 nm [86]. Thus, the design of the filter with a pore size of less than 100 nm can be a more effective strategy to remove SARS-CoV-2 from water. As we already discussed that graphene foam is widely used for the removal of toxic pollutants, dyes, oils, heavy metal ions, and microorganisms from water, therefore, graphene foam can be useful for the removal of SARS-CoV-2 from water. In addition, graphene foam is a porous material with a pore range from nano to micron size which provides high surface area and antibacterial properties for the removal of the virus. Thus, a more hydrophobic nature and pore size <100 nm make graphene foam a potential candidate for the removal of SARS-CoV-2 from wastewater [87]. Palmieri et al. [88] published a review article on GO membrane for the filtration of SARS-CoV-2. Similarly, Nasir and his coworkers [89] reported a recent review on the membrane technology for the separation of SARS-CoV-2 from wastewater. Till now, very limited work has been carried out for the removal of SARS-CoV-2 from wastewater using graphene foam. Therefore, we should focus on the design and fabrication of graphene foam for the filtration of SARS-CoV-2 from wastewater in the future.

7.4 CONCLUSION AND FUTURE PERSPECTIVE

In this chapter, we reviewed the current progress of graphene foam synthesis for biosensors, biomedical applications, removal of toxic metals, organic pollutants, and coronavirus from wastewater. Several methods and strategies to develop light-weight graphene foams have been reviewed. Also, graphene foams are modified with various metal nanoparticles and metal oxide nanoparticles, and the role of these nanoparticles on the properties of graphene foam has been discussed. The role of porosity and surface area in graphene foam is found to be a benefit in the biosensors and removal of toxic pollutants and coronavirus from contaminated water. Such graphene foam–based filters could be used in air purification to efficiently filter COVID-19 and other viruses present in the air. In addition, these graphene foams can be used to separate COVID-19 viruses from water by tuning their pore size and surface area. Instead of graphene foam, the newly developed 2D materials like MXene and Borophene-based foam or aerogel could be useful for multifunctional applications.

REFERENCES

1. Geim AK, Novoselov KS. The rise of graphene. In: Nanoscience and technology: a collection of reviews from nature journals: World Scientific; 2010:11–9.
2. Stoller MD, Park S, Zhu Y, An J, Ruoff RS. Graphene-based ultracapacitors. Nano Letters. 2008;8:3498–502.
3. Lee C, Wei X, Kysar JW, Hone J. Measurement of the elastic properties and intrinsic strength of monolayer graphene. Science. 2008;321:385–8.
4. Balandin AA, Ghosh S, Bao W, Calizo I, Teweldebrhan D, Miao F, Lau CN. Superior thermal conductivity of single-layer graphene. Nano Letters. 2008;8:902–7.
5. Sang M, Shin J, Kim K, Yu KJ. Electronic and thermal properties of graphene and recent advances in graphene based electronics applications. Nanomaterials. 2019;9:374.
6. Chang H, Wu H. Graphene-based nanomaterials: synthesis, properties, and optical and optoelectronic applications. Advanced Functional Materials. 2013;23:1984–97.
7. Olabi A, Abdelkareem MA, Wilberforce T, Sayed ET. Application of graphene in energy storage device—a review. Renewable and Sustainable Energy Reviews. 2021;135:110026.
8. Song S, Shen H, Wang Y, Chu X, Xie J, Zhou N, Shen J. Biomedical application of graphene: from drug delivery, tumor therapy, to theranostics. Colloids and Surfaces B: Biointerfaces. 2020;185:110596.
9. Prattis I, Hui E, Gubeljak P, Schierle GSK, Lombardo A, Occhipinti LG. Graphene for biosensing applications in point-of-care testing. Trends in Biotechnology. 2021;39:1065–77.
10. Amani H, Mostafavi E, Arzaghi H, Davaran S, Akbarzadeh A, Akhavan O, Pazoki-Toroudi H, Webster TJ. Three-dimensional graphene foams: synthesis, properties, biocompatibility, biodegradability, and applications in tissue engineering. ACS Biomaterials Science & Engineering. 2018;5:193–214.
11. Jiang L, Fan Z. Design of advanced porous graphene materials: from graphene nanomesh to 3D architectures. Nanoscale. 2014;6:1922–45.
12. Xu X, Guan C, Xu L, Tan YH, Zhang D, Wang Y, Zhang H, Blackwood DJ, Wang J, Li M, Ding J. Three dimensionally free-formable graphene foam with designed structures for energy and environmental applications. ACS Nano. 2020;14:937–47.
13. Chen Z, Ren W, Gao L, Liu B, Pei S, Cheng H-M. Three-dimensional flexible and conductive interconnected graphene networks grown by chemical vapour deposition. Nature Materials. 2011;10:424–8.

14. Chen Z, Xu C, Ma C, Ren W, Cheng H-M. Lightweight and flexible graphene foam composites for high-performance electromagnetic interference shielding. Advanced Materials. 2013;25:1296–300.
15. Dong X, Wang X, Wang L, Song H, Zhang H, Huang W, Chen P. 3D graphene foam as a monolithic and macroporous carbon electrode for electrochemical sensing. ACS Applied Materials & Interfaces. 2012;4:3129–33.
16. Sun Y, Wang C, Xue Y, Zhang Q, Mendes RG, Chen L, Zhang T, Gemming T, Rümmeli MH, Ai X. Coral-inspired nanoengineering design for long-cycle and flexible lithium-ion battery anode. ACS Applied Materials & Interfaces. 2016;8:9185–93.
17. Shi L, Chen K, Du R, Bachmatiuk A, Rümmeli MH, Xie K, Huang Y, Zhang Y, Liu Z. Scalable seashell-based chemical vapor deposition growth of three-dimensional graphene foams for oil—water separation. Journal of the American Chemical Society. 2016;138:6360–3.
18. Samad YA, Li Y, Alhassan SM, Liao K. Novel graphene foam composite with adjustable sensitivity for sensor applications. ACS Applied Materials & Interfaces. 2015;7:9195–202.
19. Shen B, Li Y, Zhai W, Zheng W. Compressible graphene-coated polymer foams with ultralow density for adjustable electromagnetic interference (EMI) shielding. ACS Applied Materials & Interfaces. 2016;8:8050–7.
20. Yadav A, Kumar R, Bhatia G, Verma G. Development of mesophase pitch derived high thermal conductivity graphite foam using a template method. Carbon. 2011;49:3622–30.
21. Yuan Y, Liu L, Yang M, Zhang T, Xu F, Lin Z, Ding Y, Wang C, Li J, Yin W. Lightweight, thermally insulating and stiff carbon honeycomb-induced graphene composite foams with a horizontal laminated structure for electromagnetic interference shielding. Carbon. 2017;123:223–32.
22. Subrati A, Mondal S, Ali M, Alhindi A, Ghazi R, Abdala A, Reinalda D, Alhassan S. Developing hydrophobic graphene foam for oil spill cleanup. Industrial & Engineering Chemistry Research. 2017;56:6945–51.
23. Deng W, Fang Q, Zhou X, Cao H, Liu Z. Hydrothermal self-assembly of graphene foams with controllable pore size. RSC Advances. 2016;6:20843–9.
24. Ji C-C, Xu M-W, Bao S-J, Cai C-J, Lu Z-J, Chai H, Yang F, Wei H. Self-assembly of three-dimensional interconnected graphene-based aerogels and its application in super-capacitors. Journal of Colloid and Interface Science. 2013;407:416–24.
25. Ma Y, Yue Y, Zhang H, Cheng F, Zhao W, Rao J, Luo S, Wang J, Jiang X, Liu Z. 3D syn-ergistical MXene/reduced graphene oxide aerogel for a piezoresistive sensor. ACS Nano. 2018;12:3209–16.
26. Dai Y, Wu X, Liu Z, Zhang H-B, Yu Z-Z. Highly sensitive, robust and anisotropic MXene aerogels for efficient broadband microwave absorption. Composites Part B: Engineering. 2020;200:108263.
27. Wang X, Zhang Y, Zhi C, Wang X, Tang D, Xu Y, Weng Q, Jiang X, Mitome M, Golberg D. Three-dimensional strutted graphene grown by substrate-free sugar blowing for high-power-density supercapacitors. Nature Communications. 2013;4:1–8.
28. Han F, Qian O, Chen B, Tang H, Wang M. Sugar blowing-assisted reduction and inter-connection of graphene oxide into three-dimensional porous graphene. Journal of Alloys and Compounds. 2018;730:386–91.
29. Wang Y, Huang J, Chen X, Wang L, Ye Z. Powder metallurgy template growth of 3D N-doped graphene foam as binder-free cathode for high-performance lithium/sulfur bat-tery. Carbon. 2018;137:368–78.
30. Sha J, Gao C, Lee S-K, Li Y, Zhao N, Tour JM. Preparation of three-dimensional gra-phene foams using powder metallurgy templates. ACS Nano. 2016;10:1411–6.
31. Huang X, Qian K, Yang J, Zhang J, Li L, Yu C, Zhao D. Functional nanoporous graphene foams with controlled pore sizes. Advanced Materials. 2012;24:4419–23.

32. Sha J, Li Y, Villegas Salvatierra R, Wang T, Dong P, Ji Y, Lee S-K, Zhang C, Zhang J, Smith RH. Three-dimensional printed graphene foams. ACS Nano. 2017;11:6860–7.
33. Joshi P, Mishra R, Narayan RJ. Biosensing applications of carbon-based materials. Current Opinion in Biomedical Engineering. 2021;18:100274.
34. Qiu H-J, Guan Y, Luo P, Wang Y. Recent advance in fabricating monolithic 3D porous graphene and their applications in biosensing and biofuel cells. Biosensors and Bioelectronics. 2017;89:85–95.
35. Yue HY, Zhang H, Chang J, Gao X, Huang S, Yao LH, Lin XY, Guo EJ. Highly sensitive and selective uric acid biosensor based on a three-dimensional graphene foam/indium tin oxide glass electrode. Analytical Biochemistry. 2015;488:22–7.
36. Guo J, Zhang T, Hu C, Fu L. A three-dimensional nitrogen-doped graphene structure: a highly efficient carrier of enzymes for biosensors. Nanoscale. 2015;7:1290–5.
37. Wang Y, Li H, Kong J. Facile preparation of mesocellular graphene foam for direct glucose oxidase electrochemistry and sensitive glucose sensing. Sensors and Actuators B: Chemical. 2014;193:708–14.
38. Wang Y-H, Huang K-J, Wu X. Recent advances in transition-metal dichalcogenides based electrochemical biosensors: a review. Biosensors and Bioelectronics. 2017;97:305–16.
39. Kung C-C, Lin P-Y, Buse FJ, Xue Y, Yu X, Dai L, Liu C-C. Preparation and characterization of three dimensional graphene foam supported platinum—ruthenium bimetallic nanocatalysts for hydrogen peroxide based electrochemical biosensors. Biosensors and Bioelectronics. 2014;52:1–7.
40. Yuan C-X, Fan Y-R, Guo H-X, Zhang J-X, Wang Y-L, Shan D-L, Lu X-Q. A new electrochemical sensor of nitro aromatic compound based on three-dimensional porous Pt-Pd nanoparticles supported by graphene—multiwalled carbon nanotube composite. Biosensors and Bioelectronics. 2014;58:85–91.
41. Huang B, Liu J, Lai L, Yu F, Ying X, Ye B-C, Li Y. A free-standing electrochemical sensor based on graphene foam-carbon nanotube composite coupled with gold nanoparticles and its sensing application for electrochemical determination of dopamine and uric acid. Journal of Electroanalytical Chemistry. 2017;801:129–34.
42. Yang C, Xu Y, Man P, Zhang H, Huo Y, Yang C, Li Z, Jiang S, Man B. Formation of large-area stretchable 3D graphene—nickel particle foams and their sensor applications. RSC Advances. 2017;7:35016–26.
43. Fazio E, Spadaro S, Corsaro C, Neri G, Leonardi SG, Neri F, Lavanya N, Sekar C, Donato N, Neri G. Metal-oxide based nanomaterials: synthesis, characterization and their applications in electrical and electrochemical sensors. Sensors. 2021;21:2494.
44. Dong X-C, Xu H, Wang X-W, Huang Y-X, Chan-Park MB, Zhang H, Wang L-H, Huang W, Chen P. 3D graphene—cobalt oxide electrode for high-performance supercapacitor and enzymeless glucose detection. ACS Nano. 2012;6:3206–13.
45. Si P, Dong X-C, Chen P, Kim D-H. A hierarchically structured composite of Mn_3O_4/3D graphene foam for flexible nonenzymatic biosensors. Journal of Materials Chemistry B. 2013;1:110–5.
46. Yue HY, Huang S, Chang J, Heo C, Yao F, Adhikari S, Gunes F, Liu LC, Lee TH, Oh ES. ZnO nanowire arrays on 3D hierachical graphene foam: biomarker detection of Parkinson's disease. ACS Nano. 2014;8:1639–46.
47. Yue HY, Song SS, Guo XR, Huang S, Gao X, Wang Z, Wang WQ, Zhang HJ, Wu PF. Three-dimensional ZnO nanosheet spheres/graphene foam for electrochemical determination of levodopa in the presence of uric acid. Journal of Electroanalytical Chemistry. 2019;838:142–7.
48. Yue HY, Zhang HJ, Huang S, Lu XX, Gao X, Song SS, Wang Z, Wang WQ, Guan EH. Highly sensitive and selective dopamine biosensor using Au nanoparticles-ZnO nanocone arrays/graphene foam electrode. Materials Science and Engineering: C. 2020;108:110490.

49. Güneş F, Aykaç A, Erol M, Erdem Ç, Hano H, Uzunbayir B, Şen M, Erdem A. Synthesis of hierarchical hetero-composite of graphene foam/α-Fe$_2$O$_3$ nanowires and its application on glucose biosensors. Journal of Alloys and Compounds. 2022;895:162688.
50. Krueger E, Chang AN, Brown D, Eixenberger J, Brown R, Rastegar S, Yocham KM, Cantley KD, Estrada D. Graphene foam as a three-dimensional platform for myotube growth. ACS Biomaterials Science & Engineering. 2016;2:1234–41.
51. Rajendran S, Priya TAK, Khoo KS, Hoang TKA, Ng H-S, Munawaroh HSH, Karaman C, Orooji Y, Show P+L. A critical review on various remediation approaches for heavy metal contaminants removal from contaminated soils. Chemosphere. 2022;287:132369.
52. Sajid M, Nazal MK, Ihsanullah, Baig N, Osman AM. Removal of heavy metals and organic pollutants from water using dendritic polymers based adsorbents: a critical review. Separation and Purification Technology. 2018;191:400–23.
53. Zhang Y, Zhu C, Liu F, Yuan Y, Wu H, Li A. Effects of ionic strength on removal of toxic pollutants from aqueous media with multifarious adsorbents: a review. Science of the Total Environment. 2019;646:265–79.
54. Chowdhury S, Balasubramanian R. Recent advances in the use of graphene-family nano-adsorbents for removal of toxic pollutants from wastewater. Advances in Colloid and Interface Science. 2014;204:35–56.
55. Thakur K, Kandasubramanian B. Graphene and graphene oxide-based composites for removal of organic pollutants: a review. Journal of Chemical & Engineering Data. 2019;64:833–67.
56. Wang H, Mi X, Li Y, Zhan S. 3D graphene-based macrostructures for water treatment. Advanced Materials. 2020;32:1806843.
57. Yap PL, Nine MJ, Hassan K, Tung TT, Tran DNH, Losic D. Graphene-based sorbents for multipollutants removal in water: a review of recent progress. Advanced Functional Materials. 2021;31:2007356.
58. Remanan S, Padmavathy N, Ghosh S, Mondal S, Bose S, Das NC. Porous graphene-based membranes: preparation and properties of a unique two-dimensional nanomaterial membrane for water purification. Separation & Purification Reviews. 2021;50:262–82.
59. Zhang X, Liu D, Yang L, Zhou L, You T. Self-assembled three-dimensional graphene-based materials for dye adsorption and catalysis. Journal of Materials Chemistry A. 2015;3:10031–7.
60. Mei J-Y, Qi P, Wei X-N, Zheng X-C, Wang Q, Guan X-X. Assembly and enhanced elimination performance of 3D graphene aerogel-zinc oxide hybrids for methylene blue dye in water. Materials Research Bulletin. 2019;109:141–8.
61. Xu J, Du P, Bi W, Yao G, Li S, Liu H. Graphene oxide aerogels co-functionalized with polydopamine and polyethylenimine for the adsorption of anionic dyes and organic solvents. Chemical Engineering Research and Design. 2020;154:192–202.
62. Liu H, Tian X, Xiang X, Chen S. Preparation of carboxymethyl cellulose/graphene composite aerogel beads and their adsorption for methylene blue. International Journal of Biological Macromolecules. 2022;202:632–43.
63. Weng D, Song L, Li W, Yan J, Chen L, Liu Y. Review on synthesis of three-dimensional graphene skeletons and their absorption performance for oily wastewater. Environmental Science and Pollution Research. 2021;28:16–34.
64. Bong J, Lim T, Seo K, Kwon C-A, Park JH, Kwak SK, Ju S. Dynamic graphene filters for selective gas-water-oil separation. Scientific Reports. 2015;5:1–6.
65. Sun H, Xu Z, Gao C. Multifunctional, ultra-flyweight, synergistically assembled carbon aerogels. Advanced Materials. 2013;25:2554–60.
66. Yang S, Chen L, Mu L, Hao B, Chen J, Ma P-C. Graphene foam with hierarchical structures for the removal of organic pollutants from water. RSC Advances. 2016;6:4889–98.
67. Krebsz M, Pasinszki T, Tung TT, Nine MJ, Losic D. Multiple applications of bio-graphene foam for efficient chromate ion removal and oil-water separation. Chemosphere. 2021;263:127790.

68. Mi X, Huang G, Xie W, Wang W, Liu Y, Gao J. Preparation of graphene oxide aerogel and its adsorption for Cu2+ ions. Carbon. 2012;50:4856–64.
69. Qin SY, Liu XJ, Zhuo RX, Zhang XZ. Microstructure-controllable graphene oxide hydrogel film based on a pH-responsive graphene oxide hydrogel. Macromolecular Chemistry and Physics. 2012;213:2044–51.
70. Kabiri S, Tran DN, Azari S, Losic D. Graphene-diatom silica aerogels for efficient removal of mercury ions from water. ACS Applied Materials & Interfaces. 2015;7:11815–23.
71. Singh DK, Kumar V, Singh VK, Hasan SH. Modeling of adsorption behavior of the amine-rich GOPEI aerogel for the removal of As (III) and As (V) from aqueous media. RSC Advances. 2016;6:56684–97.
72. Gao C, Dong Z, Hao X, Yao Y, Guo S. Preparation of reduced graphene oxide aerogel and its adsorption for Pb (II). ACS Omega. 2020;5:9903–11.
73. Zhu Y-H, Zhang Q, Sun G-T, Chen C-Z, Zhu M-Q, Huang X-H. The synthesis of tannin-based graphene aerogel by hydrothermal treatment for removal of heavy metal ions. Industrial Crops and Products. 2022;176:114304.
74. Liao Y, Wang M, Chen D. Preparation of polydopamine-modified graphene oxide/chitosan aerogel for uranium(VI) adsorption. Industrial & Engineering Chemistry Research. 2018;57:8472–83.
75. Deng X, Liu X, Duan T, Zhu W, Yi Z, Yao W. Fabricating a graphene oxide—bayberry tannin sponge for effective radionuclide removal. Materials Research Express. 2016;3:055002.
76. Jang S-C, Haldorai Y, Lee G-W, Hwang S-K, Han Y-K, Roh C, Huh YS. Porous three-dimensional graphene foam/Prussian blue composite for efficient removal of radioactive 137Cs. Scientific Reports. 2015;5:17510.
77. Li Y, Li L, Chen T, Duan T, Yao W, Zheng K, Dai L, Zhu W. Bioassembly of fungal hypha/graphene oxide aerogel as high performance adsorbents for U(VI) removal. Chemical Engineering Journal. 2018;347:407–14.
78. Chen J, Lu H, Melino G, Boccia S, Piacentini M, Ricciardi W, Wang Y, Shi Y, Zhu T. COVID-19 infection: the China and Italy perspectives. Cell Death & Disease. 2020;11:1–17.
79. Li Q, Guan X, Wu P, Wang X, Zhou L, Tong Y, Ren R, Leung KS, Lau EH, Wong JY. Early transmission dynamics in Wuhan, China, of novel coronavirus—infected pneumonia. New England Journal of Medicine. 2020;382:1199–207.
80. Usman M, Farooq M, Hanna K. Existence of SARS-CoV-2 in wastewater: implications for its environmental transmission in developing communities. Environmental Science & Technology. 2020;54:7758–9.
81. Kweinor Tetteh E, Opoku Amankwa M, Armah EK, Rathilal S. Fate of covid-19 occurrences in wastewater systems: emerging detection and treatment technologies—a review. Water. 2020;12:2680.
82. Ling Y, Xu S-B, Lin Y-X, Tian D, Zhu Z-Q, Dai F-H, Wu F, Song Z-G, Huang W, Chen J. Persistence and clearance of viral RNA in 2019 novel coronavirus disease rehabilitation patients. Chinese Medical Journal. 2020;133:1039–43.
83. Bansiddhi A, Dunand DC. Processing of NiTi foams by transient liquid phase sintering. Journal of Materials Engineering and Performance. 2011;20:511–6.
84. Ihsanullah I, Bilal M, Naushad M. Coronavirus 2 (SARS-CoV-2) in water environments: current status, challenges and research opportunities. Journal of Water Process Engineering. 2021;39:101735.
85. Adelodun B, Ajibade FO, Ibrahim RG, Bakare HO, Choi K-S. Snowballing transmission of COVID-19 (SARS-CoV-2) through wastewater: any sustainable preventive measures to curtail the scourge in low-income countries? Science of the Total Environment. 2020;742:140680.
86. Zhu N, Zhang D, Wang W, Li X, Yang B, Song J, Zhao X, Huang B, Shi W, Lu R. A novel coronavirus from patients with pneumonia in China, 2019. New England Journal of Medicine. 2020;382:727–33.

87. Afroj S, Britnell L, Hasan T, Andreeva DV, Novoselov KS, Karim N. Graphene-based technologies for tackling COVID-19 and future pandemics. Advanced Functional Materials. 2021:2107407.

88. Palmieri V, Papi M. Can graphene take part in the fight against COVID-19? Nano Today. 2020:100883.

89. Nasir AM, Adam MR, Kamal SNEAM, Jaafar J, Othman MHD, Ismail AF, Aziz F, Yusof N, Bilad MR, Mohamud R. A review of the potential of conventional and advanced membrane technology in the removal of pathogens from wastewater. Separation and Purification Technology. 2022:120454.

8 2D Materials–Based High-Performance Sterilizers for COVID-19

Neeraj Dwivedi

A *sterilizer* is a device or a system that destroys viruses, bacteria, or any other microorganism by means of dry heat, steam, or ultraviolet (UV) radiation. Due to very sensitive environment, sterilizers are very popular in hospitals. For example, the components that are used for surgical/non-surgical functions and come in contact with the humans must be sterilized before their use. Otherwise, there could be high chances of bacterial, viral, or other microorganism transfer to the person. This process is called contact transmission. Another classic example of a sterilizer is in "the sterilization of milk bottles of children using steam or any other method in order to prevent the children from infectious microorganisms." Thus, sterilizers are a crucial component for controlling the transmission of microorganisms and minimizing the risk of infection spread, including COVID-19.

Sterilizers have been in use since long back [1]. Pitch and tar were used as antiseptics by the Egyptians in 3000 BC. Then, sulfur fumes were used to disinfect systems. After the discovery of pressure cooker in 1680, which converts boiling water into steam, steam-based method became popular for sterilization. With improved design of steam sterilizers, it was possible to sterilize solid objects, including textiles, utensils, etc. Hence, in the 1800s, steam-based sterilizers were very popular. In the 1860s, Joseph Lister employed carbolic acid–based spray to disinfect the objects. More innovations in sterilizations happened in 1900, where, along with steam, sterilization by radiation and glutaraldehyde was also started. Of late, sterilization by high-temperature system, UV radiation, and various chemicals is common and in use. Since most widely used sterilizers in the present days are based on UV radiation and they have been widely used during COVID-19, we will first discuss more about UV-based sterilizers.

A COVID-19-infected person, when coughing, can generate few thousands of droplets, which then deposit onto the surfaces [2]. Through high-touch surfaces, such as railings, door handles, window glasses, utensils, and many more, the SARS-CoV-2 virus can be transmitted to humans. Importantly, it is reported that the SARS-CoV-2 virus can survive from a few hours to a few days, depending upon the type of material, thus raising high chances of infection transmission [3, 4]. UV-based disinfectant systems are known for disinfection of microorganisms on surfaces and various other mediums, including water and air. While the UV light range is quite large, the most commonly used wavelength range for UV-based disinfectants is 200–280

 DOI: 10.1201/9781003316381-8

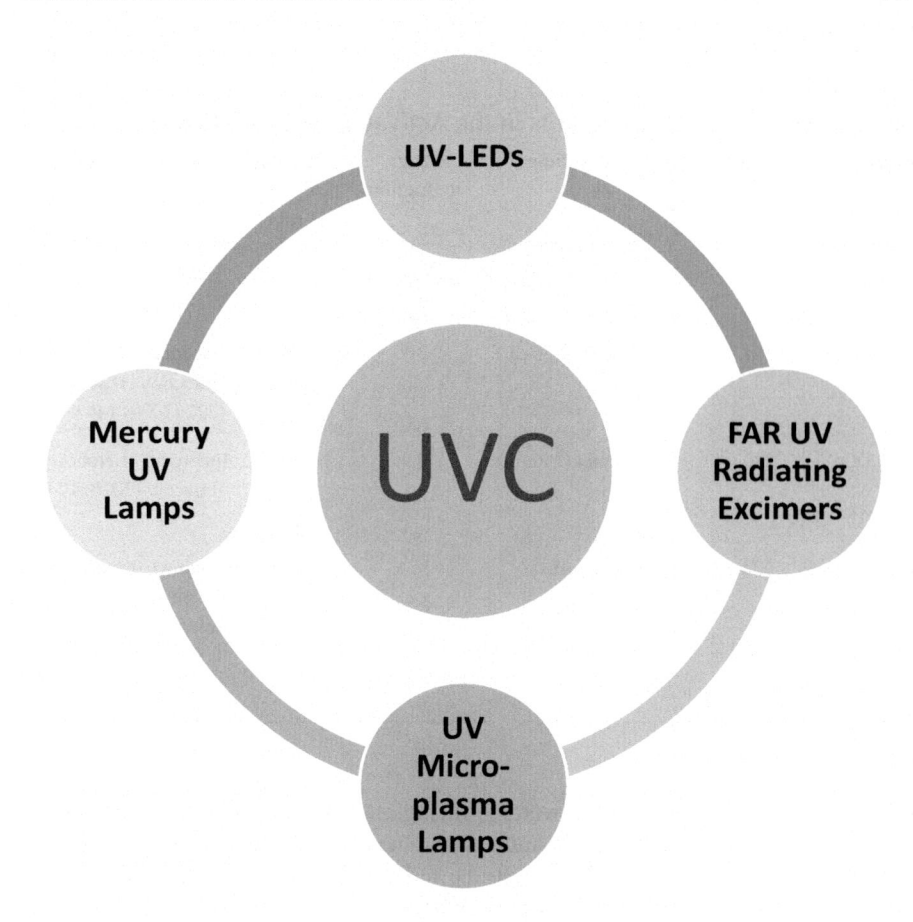

FIGURE 8.1   Schematic illustration of different UVC light sources.

nm, also called the UVC spectrum range [2]. When microorganisms such as viruses or bacteria are exposed to UVC light, they absorb the light due to the interaction of their molecular structure with light, which then destroys the genomic structure of microorganisms and lead to the inability of the cells to replicate. To accomplish the UVC disinfectants, the UVC light source is a major component. A wide variety of UVC light sources is available [2], as shown in Figure 8.1. Different studies have been performed to explore the UVC light of different wavelengths for inactivation of SARS-CoV-2; some of them are listed in Table 8.1.

Smart disinfectants can also be developed by exploiting the photothermal and photocatalytic properties of the materials. However, the fundamental requirement for such systems is that the materials should hold the promising photo-to-heat conversion and photocatalytic characteristics. The photothermal property is associated with electromagnetic radiation, where the electromagnetic radiation gets converted into thermal energy *via* materials-mediated approach. This concept is also used in photothermal therapy. The photocatalytic phenomenon exploits the synergistic effect of photons and catalysts that boost the chemical reactions.

TABLE 8.1

Application of UVC Disinfectants in the Activation of SARS-CoV-2

Virus	Wavelength (nm)	Medium	Log Reduction	UV Dose (mj/cm² or min.)	Remarks
SARS-CoV-2	254	Liquid (976 µL)	3	3.7	Sci. Rep. 11 (2021) 6260
			6	16.9	
			6	84.4	
SARS-CoV-2	254	Liquid (250 µL)	2.1	10	The Journal of Infectious Diseases 222 (2020) 1462
			3.9	20	
			6.0	40	
SARS-CoV-2	280±5	Liquid (150 µL)	0.9	3.75	Emerging Microbes Infect. 9 (2020) 1744
			3.1	37.5	
			3.3	75	
SARS-CoV-2	200–320	Hard surface	3.56	1 min	Infection Control & Hospital Epidemiology 42 (2021) 127
			4.54	2 min	
			4.12	5 min	

Source: Reproduced from refs. [2, 5–8].

Photocatalysis is widely used in water splitting and various other electrochemical reactions. Emerging 2D materials, such as graphene, graphene oxide (GO), reduced graphene oxide (rGO), and their hybrids and composites with other materials, possess excellent photothermal properties. Researchers have exploited structured graphene metamaterials for light absorption and thermal energy conversion [9]. Tang et al. [10] developed paraffin/graphene aerogel shape-stable phase-change materials (SSPCMs) for light-to-heat conversion application. The schematic illustration for preparation of SSPCMs is presented in Figure 8.2. Paraffin/graphene aerogel SSPCMs display better light-to-heat conversion efficiency than pristine paraffin.

Robinson et al. [11] developed nano-graphene oxide (nano-GO) and nano-reduced graphene oxide (nano-rGO) for light-to-heat conversion of near-infrared light for photothermal therapy application. Figure 8.3a shows nano-GO- and nano-rGO-containing solutions. It is evident from Figure 8.3b that nano-rGO shows significantly higher near-infrared light absorption (beyond 750 nm wavelength) than nano-GO, which is remarkable. The higher light absorption in nano-rGO contributes to its higher light-to-heat conversion capability than nano-GO, which is confirmed by observation of higher temperature rise in nano-rGO (Figure 8.3c) than nano-GO (Figure 8.3d). Zhuang et al. prepared rGO for photo-to-thermal conversion application and realized up to 79% efficiency for conversion of solar light to thermal energy [12].

MoS_2 is another interesting 2D material that displays excellent photo-to-heat conversion efficiency. Salimi et al. investigated the photo-to-heat conversion ability of

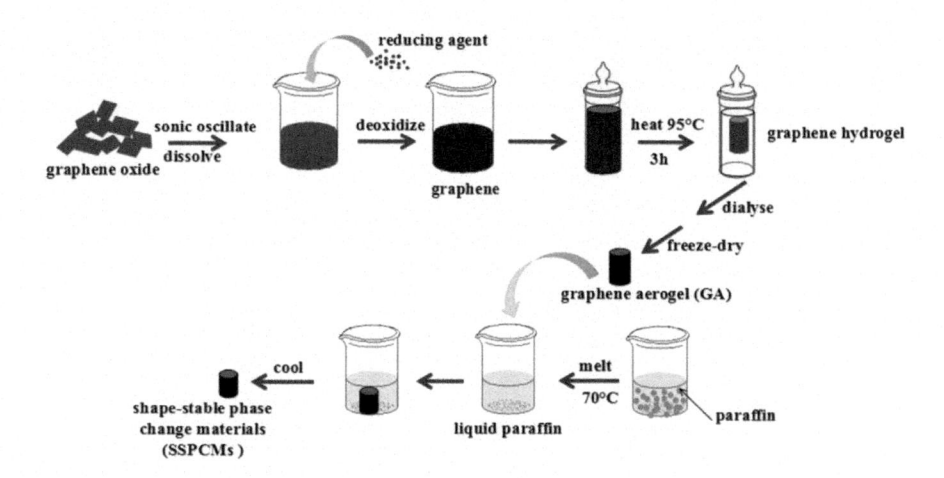

FIGURE 8.2 Schematic illustration for preparation of paraffin/graphene aerogel SSPCMs.

Source: Reproduced from ref. [10].

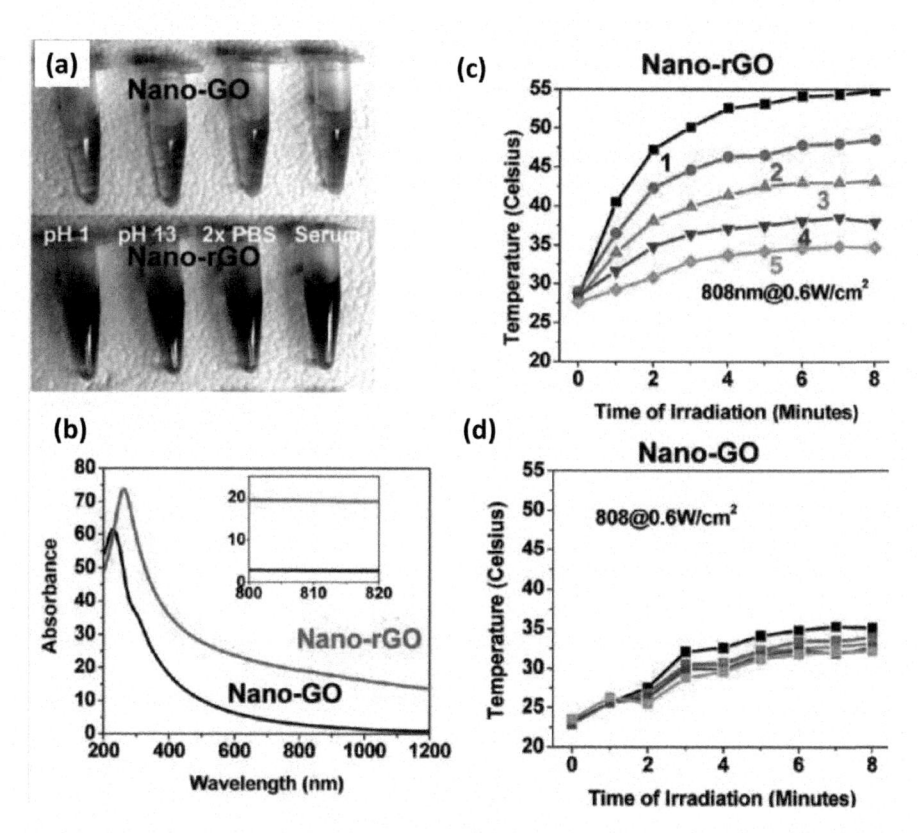

FIGURE 8.3 (a) Nano-GO- and nano-rGO-containing solutions; (b) absorbance spectra and temperature profile as a function of NIR irradiation time for (c) nano-rGO and (d) nano-GO samples.

Source: Reproduced with permission from ref. [11].

two types of MoS_2 structures: (1) MoS_2 nanosheet (MoS_2-NS) and MoS_2 nanoflowers (MoS_2-NFs) [13]. The optical results revealed the higher NIR light absorption in MoS_2-NS sample than MoS_2-NFs. The quantitative estimation of photo-to-heat conversion efficiency revealed higher conversion efficiency of 25.7% in MoS_2-NS than 13.8% in MoS_2-NFs, which was mainly attributed to the presence of higher 1T/2H ratio in MoS_2-NS sample [13]. Abareshi et al. investigated the photothermal property of MoS_2 nanoflakes produced by hydrothermal method [14]. The sample was exposed to NIR 808 nm radiation, resulting in rise of temperature on nanoflakes due to conversion of NIR light to heat.

MXene and composites of MXene also display excellent light-to-heat conversion property. Li et al. explored Ti_3C_2 MXene as light-to-heat conversion material [15]. They started with etching of Ti_3AlC_2 MAX precursor in HF, followed by sonication-assisted exfoliation in DMSO, leading to the development of single- to few-layered Ti_3C_2 MXene. To examine the light-to-heat conversion capability of MXene, a droplet-based light absorption and heat measurement system was developed. The droplet containing MXene was suspended at one end of the PTFE tube, which was then exposed to light of known wavelength coming from a laser source to examine the optical absorption and heat generation capability of MXene [15]. The optical absorbance of the MXene-containing droplet was studied and compared with carbon nanotube (CNT) containing droplet. It was observed that MXene displays enhanced optical absorbance with distinct peak than CNT. The temperature response of the MXene-containing droplet under two different laser wavelengths clearly shows heat generation, suggesting MXene as a promising photo-to-heat conversion material [15].

Further, Li et al. [16] exploited biomimetic MXene-based textures for the conversion of light to heat. In particular, crumpled textures inspired by *Bitis rhinoceros* were designed. The application of nanocoatings of MXene-based materials helped boost the light-to-heat conversion capability of crumpled textures due to enhanced absorbance with respect to uncoated textures. Huang et al. [17] stressed that MXene-based materials are excellent photo-to-heat conversion agents. Yang et al. [18] developed Au nanorods/V_2C MXene membranes for efficient light-to-heat conversion application. The synergistic effects of Au nanorods and V_2C MXene boost the light absorption and, hence, enhance photo-to-heat conversion.

The exciting photothermal properties of graphene-, GRM-, and MXene-based materials can potentially be exploited to develop sterilizers. These 2D materials in sterilizers would act as light-to-heat converters, and the contaminated objects placed in sterilizers will be disinfected, that is, the microorganisms will be inactivated due to their exposure to heat. Furthermore, MXene also possesses hydrophilic characteristic [19], which can be harnessed in developing efficient sterilizers too. The synergism of light-to-heat conversion and hydrophilic characteristics of MXenes can be exploited to inactivate the microbes by the following ways: (a) The sharp edges of MXenes would rupture the cell membranes of microorganisms due to intimate contacts of material and contaminants. (b) The light-to-heat conversion efficacy of MXene materials will enable fast transfer of heat across the intimate contacted microorganisms, which facilitates faster deactivation of living contaminants. Thus,

MXene-based materials can be explored for the development of photo-sterilizers. The hybrids or composites of graphene-MXene or GRMs-MXene can show even better photo-sterilizing effects than graphene, GRMs, or MXenes standalone due to synergistic effects in the former.

Next, photocatalysts can also be exploited to develop sterilizers. Photocatalysts are materials which boost chemical reactions *via* light-induced effects. Semiconducting materials are mainly explored as photocatalysts. The photocatalytic process includes (a) light absorption by photocatalysts and, subsequently, the generation of electron-hole pairs, (b) separation of charge carriers and their transfer, and (c) surface reactions [20, 21]. Traditional photocatalysts, such as metal oxides, especially TiO_2, g-C_3N_4, CdS, ZnO, etc., have certain drawbacks, mainly, the early recombination of photo-generated charge carriers [22]. Many metals, such as Pt, Pd, Ru, etc., are employed as co-catalysts to overcome charge recombination concern and improving the charge separation, but they are extremely costly [22]. Graphene- and GRM-based materials have shown huge potential as photocatalysts, either as primary catalyst or as co-catalyst material. Graphene possesses high surface area, displays excellent electrical and thermal conductivities, and can be functionalized and integrated with various other material systems [23]. Li et al. extensively discussed the role of graphene-based materials for the development of high-performance heterogenous photocatalysts [24]. It was suggested that graphene-based materials can be employed for fast charge transfer as well as photosensitizer in a heterogenous photocatalyst. While studying the GO/TiO_2 photocatalyst, Zhou et al. found that the application of GO boosts the photocatalytic activity of TiO_2 [25]. Janczarek et al. [26] critically reviewed the role of graphene, GO, rGO, and their composites as photocatalysts for various applications.

Al-Rawashdeh et al. worked on developing GO/ZnO nanocomposites having embedded metal nanoparticles for photocatalytic application [27]. They found that the GO/ZnO nanocomposite itself displays excellent photocatalytic activity, which was further tuned by embedding silver and copper nanoparticles. Researchers developed water-dispersible ZnO/$C_0Fe_2O_4$/graphene-based nanocomposites, which exhibited excellent photocatalytic activity [28]. While TiO_2 is one of the most popular photocatalysts, it suffers from poor visible light absorption due to its wide bandgap feature. Therefore, researchers have looked for narrow-bandgap semiconductors. Silver sulfide (Ag_2S) is one such promising material in this regard, and photocatalytic activity has been observed in this material. Hu et al. developed Ag_2S/graphene (GR) composites and studied photocatalytic properties in visible light region [29]. The properties of Ag_2S/GR composites were also compared with Ag_2S photocatalyst alone. The optical absorption results for Ag_2S/GR and Ag_2S samples are shown in Figure 8.4a, which reveal excellent optical absorption in both samples in visible region. However, close analysis suggested that application of graphene slightly enhances the optical absorption in Ag_2S/GR composite with respect to pristine Ag_2S sample. The photocatalytic properties also indicate enhanced photocatalytic activity in Ag_2S/GR sample than pristine Ag_2S. A model for why Ag_2S/GR composites display excellent photocatalytic properties is also proposed, as depicted in Figure 8.4b. It was suggested that the application of GR in Ag_2S/GR composite contributes to the fast charge transfer, reduction of recombination, and hence, enhancement of carrier

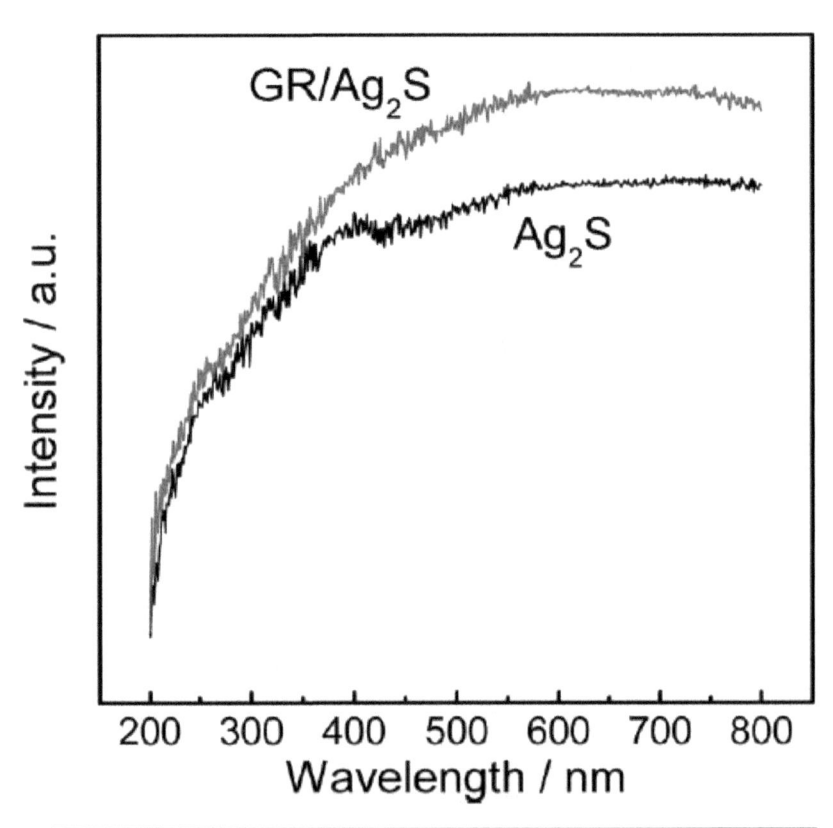

FIGURE 8.4   (a) Optical absorption results of Ag_2S and Ag_2S/GR samples, and (b) proposed model for charge separation and transportation in Ag_2S/GR composite.

Source: Reproduced from Ref. [29].

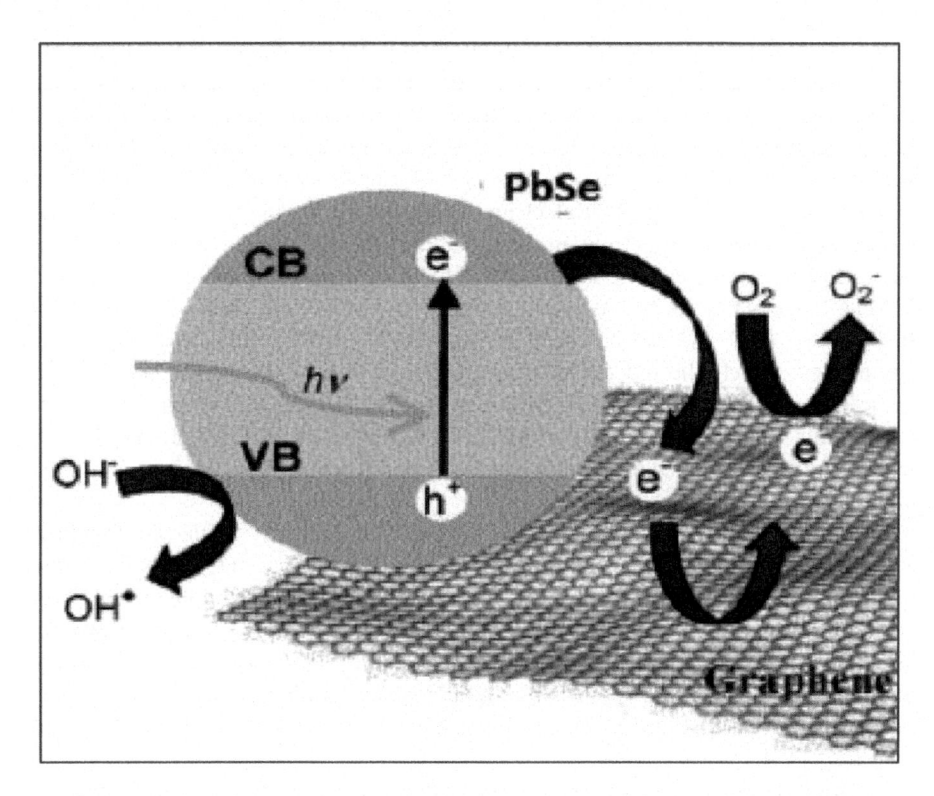

FIGURE 8.5 Schematic illustration for generation of charge carriers at PbSe/GO or graphene interface and their transportation.

Source: Reproduced from ref. [30].

lifetime, which then leads to enhanced photocatalytic activity in composite materials than pristine Ag_2S materials.

One of the interesting approaches to obtain excellent photocatalytic activity is to focus on narrow-bandgap semiconducting materials. PbSe is another narrow-bandgap semiconducting material, and its integration with graphene-based materials, such as graphene, GO, and rGO, can give rise to an outstanding photocatalytic performance. Thus, the researchers aimed to take advantage of two materials and developed PbSe/GO or graphene nanocomposites [30]. Ultrasonic-assisted method was employed to develop PbSe/GO nanocomposites. The developed PbSe/GO nanocomposites displayed improved photocatalytic activity due to high and fast charge separation induced by the synergistic effects of the two materials. The schematic illustration for charge transfer process in PbSe/GO or graphene system is presented in Figure 8.5.

MXene is another fascinating material which also displays photocatalytic activity or can be exploited as foreign materials in traditional photocatalysts to boost their photocatalytic performance [20, 22, 31, 32]. Researchers prepared MXene-based Schottky junction ($Bi_2S_3/Ti_3C_2T_x$ heterostructure) and performed work function engineering *via* interface manipulation [33]. The work function manipulation enhanced the photocatalytic performance of $Bi_2S_3/Ti_3C_2T_x$ heterostructure due to high and fast charge transfer, and hence, the heterostructure material successfully inactivated the bacteria. The MXene-based materials can be modified with surface functional groups; the engineering of functionalization of MXene-based materials can largely tune their electrical conductivity and photo-response [34, 35]. Excellent photocatalytic activity in functionalized metal carbide–based MXenes has been demonstrated [34, 35]. Further, to enhance the photocatalytic activity, Ti_3C_2 MXene has been employed as a co-catalyst on metal sulfide photo-absorbers [36]. In addition, Ti_3C_2 MXene has been incorporated in TiO_2 to construct TiO_2/Ti_3C_2 composites, which showed outstanding photocatalytic performance due to synergistic effect [37]. Further, Mo_2C MXene was also investigated for photocatalysis application. Researchers developed Mo_2C/CdS heterostructure and observed excellent photocatalytic behavior, which was exceeded to even Ti_3C_2 MXene and many other noble metal catalysts [38].

Transition metal dichalcogenides (TMDCs), especially single- to few-layer MoS_2, have emerged as a promising photocatalyst due to their suitable direct bandgap of about 1.8–1.9 eV and excellent physicochemical and electrical properties. Lai et al. developed highly expanded interlayer MoS_2 (interlayer spacing about 1.5 nm) by one-pot hydrothermal technique, which revealed outstanding photocatalytic behavior under visible light [39]. Researchers prepared TiO_2/MoS_2 heterostructure films and realized photocatalytic activity in the developed film materials [40].

Further, Ullah et al. prepared MoS_2 nanoflowers (NFs) using hydrothermal method and studied their photocatalytic property [41]. The developed MoS_2 NFs were characterized using XRD (Figure 8.6a), Raman spectroscopy (Figure 8.6b), X-ray photoelectron spectroscopy (XPS, Figure 8.6c), and FESEM (Figure 8.6d). The developed MoS_2 NFs showed excellent photocatalytic activity, which was confirmed by observing its function in degrading the Congo red (CR) dye. The possible mechanism for degradation of CR dye due to MoS_2 NFs and under the action of visible light is schematically illustrated in Figure 8.6e. Researchers employed a unique approach of preparing MoS_2 layer–anchored rGO as hybrid co-catalyst for boosting the photocatalytic performance of CdS nanorods [42]. While MOS_2 displays excellent photocatalytic property, its photocatalytic stability is also crucial for its long-term use. Parzinger et al. systematically investigated the photocatalytic stability of single- and few-layer MoS_2 obtained by exfoliation method [43]. Micro-Raman spectrophotometer was used to study its photocatalytic stability, and during the experiment, samples were immersed in water. It was observed that basal planes of MoS_2 are highly stable during photocatalytic action, while the edge sites undergo photoinduced degradation, that is, photoinduced corrosion. It was also observed that the edge sites of monolayer MoS_2 are more stable than those of few-layer MoS_2.

The photocatalytic properties of graphene, GRMs, MoS_2, MXenes, and their hybrids and composites can also enable the development of smart sterilizers for deactivating viruses and bacteria in order to combat pandemics.

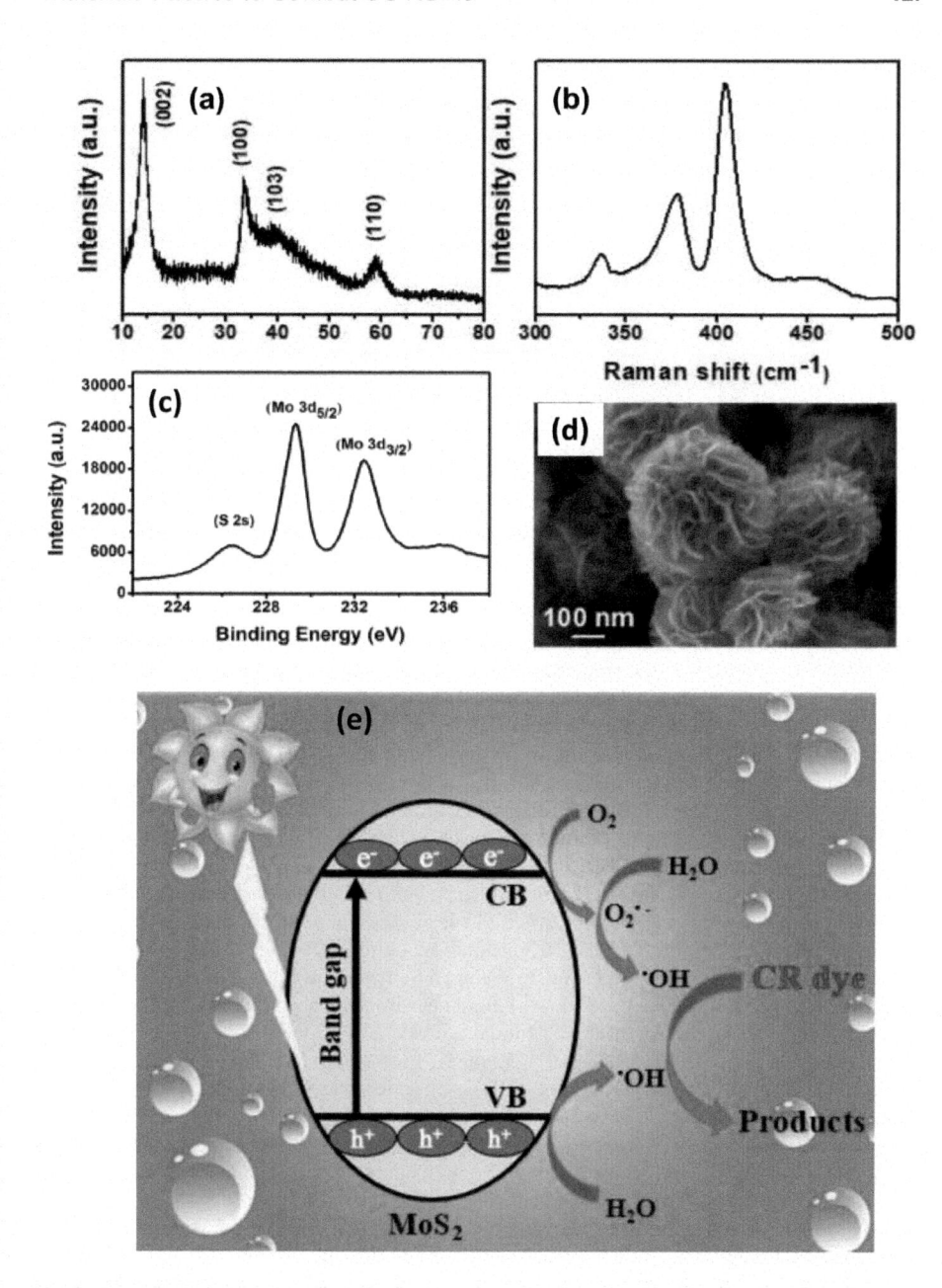

FIGURE 8.6 Characterization of MoS_2 nanoflowers (NFs) using (a) XRD, (b) Raman spectroscopy, (c) XPS (Mo 3d and S 2s spectra), and (d) FESEM; (e) how CR dye degrades due to photocatalytic action of MoS_2 NFs under visible light is schematically illustrated in the figure.

Source: Reproduced from ref. [41].

REFERENCES

1. Dwivedi N, Yeo RJ, Dhand C, Risan J, Nay R, Tripathy S, Rajauria S, Saifullah MS, Sankaranarayanan SK, Yang H. Boosting contact sliding and wear protection via atomic intermixing and tailoring of nanoscale interfaces. Science Advances. 2019;5:eaau7886.
2. Raeiszadeh M, Adeli B. A critical review on ultraviolet disinfection systems against COVID-19 outbreak: applicability, validation, and safety considerations. ACS Photonics. 2020;7:2941–51.
3. Chin AW, Chu JT, Perera MR, Hui KP, Yen H-L, Chan MC, Peiris M, Poon LL. Stability of SARS-CoV-2 in different environmental conditions. The Lancet Microbe. 2020;1:e10.
4. Van Doremalen N, Bushmaker T, Morris DH, Holbrook MG, Gamble A, Williamson BN, Tamin A, Harcourt JL, Thornburg NJ, Gerber SI. Aerosol and surface stability of SARS-CoV-2 as compared with SARS-CoV-1. New England Journal of Medicine. 2020;382:1564–7.
5. Biasin M, Bianco A, Pareschi G, Cavalleri A, Cavatorta C, Fenizia C, Galli P, Lessio L, Lualdi M, Tombetti E. UV-C irradiation is highly effective in inactivating SARS-CoV-2 replication. Scientific Reports. 2021;11:1–7.
6. Patterson EI, Prince T, Anderson ER, Casas-Sanchez A, Smith SL, Cansado-Utrilla C, Solomon T, Griffiths MJ, Acosta-Serrano Á, Turtle L. Methods of inactivation of SARS-CoV-2 for downstream biological assays. The Journal of Infectious Diseases. 2020;222:1462–7.
7. Inagaki H, Saito A, Sugiyama H, Okabayashi T, Fujimoto S. Rapid inactivation of SARS-CoV-2 with deep-UV LED irradiation. Emerging Microbes & Infections. 2020;9:1744–7.
8. Simmons SE, Carrion R, Alfson KJ, Staples HM, Jinadatha C, Jarvis WR, Sampathkumar P, Chemaly RF, Khawaja F, Povroznik M. Deactivation of SARS-CoV-2 with pulsed-xenon ultraviolet light: implications for environmental COVID-19 control. Infection Control & Hospital Epidemiology. 2021;42:127–30.
9. Lin K-T, Lin H, Yang T, Jia B. Structured graphene metamaterial selective absorbers for high efficiency and omnidirectional solar thermal energy conversion. Nature Communications. 2020;11:1–10.
10. Tang X, Luo L, Guo Y, Yang Z, Zhang K, He R, Fan J, Yang W. Preparation and light-to-heat conversion efficiency of paraffin/graphene aerogel shape-stable phase change materials. Fullerenes, Nanotubes and Carbon Nanostructures. 2019;27:375–81.
11. Robinson JT, Tabakman SM, Liang Y, Wang H, Sanchez Casalongue H, Vinh D, Dai H. Ultrasmall reduced graphene oxide with high near-infrared absorbance for photothermal therapy. Journal of the American Chemical Society. 2011;133:6825–31.
12. Zhuang P, Fu H, Xu N, Li B, Xu J, Zhou L. Free-standing reduced graphene oxide (rGO) membrane for salt-rejecting solar desalination via size effect. Nanophotonics. 2020;9:4601–8.
13. Salimi M, Shokrgozar MA, Hamid DH, Vossoughi M. Photothermal properties of two-dimensional Molybdenum Disulfide (MoS_2) with nanoflower and nanosheet morphology. Materials Research Bulletin. 2022:111837.
14. Abareshi A, Arshadi Pirlar M, Houshiar M. Photothermal property in MoS_2 nanoflakes: theoretical and experimental comparison. Materials Research Express. 2019;6:105050.
15. Li R, Zhang L, Shi L, Wang P. MXene Ti_3C_2: an effective 2D light-to-heat conversion material. ACS Nano. 2017;11:3752–9.
16. Li K, Chang TH, Li Z, Yang H, Fu F, Li T, Ho JS, Chen PY. Biomimetic MXene textures with enhanced light-to-heat conversion for solar steam generation and wearable thermal management. Advanced Energy Materials. 2019;9:1901687.
17. Huang Z, Cui X, Li S, Wei J, Li P, Wang Y, Lee C-S. Two-dimensional MXene-based materials for photothermal therapy. Nanophotonics. 2020;1.

18. Yang B, Tang P-F, Liu C-J, Li R, Li X-D, Chen J, Qiao Z-Q, Zhang H-P, Yang G-C. An efficient light-to-heat conversion coupling photothermal effect and exothermic chemical reaction in Au NRs/V_2C MXene membranes for high-performance laser ignition. Defence Technology. 2021;18:834–42.

19. Weiss C, Carriere M, Fusco L, Capua I, Regla-Nava JA, Pasquali M, Scott JA, Vitale F, Unal MA, Mattevi C. Toward nanotechnology-enabled approaches against the COVID-19 pandemic. ACS Nano. 2020;14:6383–406.

20. Kuang P, Low J, Cheng B, Yu J, Fan J. MXene-based photocatalysts. Journal of Materials Science & Technology. 2020;56:18–44.

21. Sabir R, Waheed A, Moazzam Ali M, Mushtaq U. Graphene-based photocatalysts for organic pollutant removal from waste-water: recent progress and future challenges. Environmental Technology Reviews. 2021;10:323–41.

22. Sun Y, Meng X, Dall'Agnese Y, Dall'Agnese C, Duan S, Gao Y, Chen G, Wang X-F. 2D MXenes as co-catalysts in photocatalysis: synthetic methods. Nano-Micro Letters. 2019;11:1–22.

23. Srivastava A, Dwivedi N, Dhand C, Khan R, Sathish N, Gupta MK, Kumar R, Kumar S. Potential of graphene-based materials to combat COVID-19: properties, perspectives and prospects. Materials Today Chemistry. 2020;18:100385.

24. Li X, Yu J, Wageh S, Al-Ghamdi AA, Xie J. Graphene in photocatalysis: a review. Small. 2016;12:6640–96.

25. Zhou X, Zhang X, Wang Y, Wu Z. 2D Graphene-TiO_2 composite and its photocatalytic application in water pollutants. Frontiers in Energy Research. 2021:400.

26. Janczarek M, Endo-Kimura M, Wei Z, Bielan Z, Mogan TR, Khedr TM, Wang K, Markowska-Szczupak A, Kowalska E. Novel structures and applications of graphene-based semiconductor photocatalysts: faceted particles, photonic crystals, antimicrobial and magnetic properties. Applied Sciences. 2021;11:1982.

27. Al-Rawashdeh NAF, Allabadi O, Aljarrah MT. Photocatalytic activity of graphene oxide/zinc oxide nanocomposites with embedded metal nanoparticles for the degradation of organic dyes. ACS Omega. 2020;5:28046–55.

28. Zhang L, Hu X, Zhu L, Jin X, Feng C. Water-dispersible ZnO/$COFe_2O_4$/graphene photocatalyst and their high-performance in water treatment. Fullerenes, Nanotubes and Carbon Nanostructures. 2019;27:873–7.

29. Hu W, Zhao L, Zhang Y, Zhang X, Dong L, Wang S, He Y. Preparation and photocatalytic activity of graphene-modified Ag_2S composite. Journal of Experimental Nanoscience. 2016;11:433–44.

30. Ullah K, Vikram N, Ye S, Cho K-Y, Zhu L, Meng Z-D, Oh W-C. Novel PbSe/graphene nanocomposites synthesized with ultrasonic assisted method and their enhanced photocatalytic activity. Synthesis and Reactivity in Inorganic, Metal-Organic, and Nano-Metal Chemistry. 2015;45:531–8.

31. Huang K, Li C, Li H, Ren G, Wang L, Wang W, Meng X. Photocatalytic applications of two-dimensional Ti_3C_2 MXenes: a review. ACS Applied Nano Materials. 2020;3:9581–603.

32. Tang R, Xiong S, Gong D, Deng Y, Wang Y, Su L, Ding C, Yang L, Liao C. Ti_3C_2 2D MXene: recent progress and perspectives in photocatalysis. ACS Applied Materials & Interfaces. 2020;12:56663–80.

33. Li J, Li Z, Liu X, Li C, Zheng Y, Yeung KWK, Cui Z, Liang Y, Zhu S, Hu W. Interfacial engineering of Bi_2S_3/Ti_3C_2T x MXene based on work function for rapid photo-excited bacteria-killing. Nature Communications. 2021;12:1–10.

34. Guo Z, Zhou J, Zhu L, Sun Z. MXene: a promising photocatalyst for water splitting. Journal of Materials Chemistry A. 2016;4:11446–52.

35. Wong ZM, Tan TL, Yang S-W, Xu GQ. Enhancing the photocatalytic performance of MXenes via stoichiometry engineering of their electronic and optical properties. ACS Applied Materials & Interfaces. 2018;10:39879–89.

36. Ran J, Gao G, Li F-T, Ma T-Y, Du A, Qiao S-Z. Ti_3C_2 MXene co-catalyst on metal sulfide photo-absorbers for enhanced visible-light photocatalytic hydrogen production. Nature Communications. 2017;8:1–10.

37. Low J, Zhang L, Tong T, Shen B, Yu J. TiO_2/MXene Ti_3C_2 composite with excellent photocatalytic CO_2 reduction activity. Journal of Catalysis. 2018;361:255–66.

38. Jin S, Shi Z, Jing H, Wang L, Hu Q, Chen D, Li N, Zhou A. Mo2C-MXene/CdS Heterostructures as visible-light photocatalysts with an ultrahigh hydrogen production rate. ACS Applied Energy Materials. 2021;4:12754–66.

39. Lai MTL, Lee KM, Yang TCK, Pan GT, Lai CW, Chen C-Y, Johan MR, Juan JC. The improved photocatalytic activity of highly expanded MoS_2 under visible light emitting diodes. Nanoscale Advances. 2021;3:1106–20.

40. Phung HNT, Tran VNK, Nguyen LT, Phan LKT, Duong PA, Le HVT. Investigating visible-photocatalytic activity of MoS_2/TiO_2 heterostructure thin films at various MoS_2 deposition times. Journal of Nanomaterials. 2017;2017:3197540.

41. Ullah H, Khan Z, Nasir JA, Balkan T, Butler IS, Kaya S, Zu R. Green synthesis of mesoporous MoS_2 nanoflowers for efficient photocatalytic degradation of Congo red dye. Journal of Coordination Chemistry. 2021;74:2302–14.

42. Kumar DP, Hong S, Reddy DA, Kim TK. Ultrathin MoS_2 layers anchored exfoliated reduced graphene oxide nanosheet hybrid as a highly efficient cocatalyst for CdS nanorods towards enhanced photocatalytic hydrogen production. Applied Catalysis B: Environmental. 2017;212:7–14.

43. Parzinger E, Miller B, Blaschke B, Garrido JA, Ager JW, Holleitner A, Wurstbauer U. Photocatalytic stability of single- and few-layer MoS_2. ACS Nano. 2015;9:11302–9.

9 Biocompatibility and Cytotoxicity of 2D Materials

Chetna Dhand

In ancient times, the term *biocompatibility* was not as prevalent as it is now as people utilize varieties of natural materials for different biomedical applications. The material's inherent antibacterial and cell-proliferative properties never raised any concerns about toxicity. However, the emergence of numerous man-made materials for biomedical use has led to their perception of cytotoxic effect. Several publications on the tissue response to implanted biomaterials were published in the eighteenth century. In 1970, R. J. Hegyeli and C. A. Homsey et al. appear to be the first to use the term *biocompatibility* in peer-reviewed journals [1]. One of the most common definitions was that it is a material's ability to function with an adequate host reaction in a certain application. Biocompatibility of implantable devices refers to the evaluation of substances that leak from the biomaterial or device and their impact on live cells and organisms, or the biological reaction to biomaterials after implantation in living creatures. It is the capacity of tissue engineering products to promote cell activity in order to enhance tissue regeneration while causing no damage to the host. Much work has been put into developing biocompatible materials for use in implants, antibacterials, wound treatment, and tissue engineering in recent years.

The importance of 2D materials in the biological field, as addressed in previous chapters, is widely acknowledged. Before these materials are used in clinical settings, one important attribute must be checked: their biocompatibility. Biocompatibility and cytotoxicity of many 2D materials depend on a variety of physiochemical variables, including lateral size, concentration, exposure time, number of layers, chemical composition, and surface functionalization. Using the cytotoxicity test to compare the biocompatibility of different 2D materials can help researchers better understand the mechanism behind any hazardous impact. The analysis also aids in determining the material's real-time hazard potential. In fact, prior to their possible bio applications, an in-depth understanding of their potential nanotoxicity is of utmost importance.

9.1 BIOCOMPATIBILITY OF GRAPHENE-BASED MATERIALS

Graphene-based materials can be benign or harmful to live cells, depending on their number of layers, lateral size, surface area, water retention capacity, and surface functionalization. Phagocytic and non-phagocytic cells are used to calculate nanomaterial toxicity in vitro. For medicinal applications, the interaction of these cells with

DOI: 10.1201/9781003316381-9

graphene-based material sheets is critical. Phospholipids and cholesterol molecules are present in cell membranes, where they help stabilize membrane structure, maintain fluidity, and influence membrane-related protein activities. When pure graphene comes into contact with phospholipids, cholesterol is removed from the membrane, resulting in membrane damage. According to Duan et al., GO may remove phospholipids from the cell membrane of human alveolar epithelial A549 and mouse macrophages Raw 264.7, generating a surface hole [2]. The capacity of a cell to survive is called viability. Cell viability is reduced, resulting in cell death. Strong contact and void formation were demonstrated using molecular dynamics modeling and SEM photography.

The charge and chemistry of a GO's surface have a significant influence on cellular interactions. GO has a high negative charge density due to the presence of functional groups containing oxygen, which improves the interaction between GO and membrane lipid. Li et al. investigated the GO lipid interaction using the Langmuir monolayer approach [3]. The GO interacted with positively charged lipid head groups, but not with neutral or negatively charged lipids, according to the study. As a result, cell membrane damage was induced without GO penetration into the cells. The impact of GO surface chemistry on the lipid membrane was examined using GO, hydrated GO (hGO), and rGO [4]. It has been proposed that hGO can cause lipid peroxidation of the surface membrane, resulting in membrane lysis and cell integrity degradation. The electron paramagnetic resonance (EPR) method was used to establish the production of radicals during the hydration process of GO. In human leukemic monocyte THP-1 and human bronchial epithelial BEAS-2B cells, these readily reacted with oxygen to create superoxide radicals, which may oxidize unsaturated lipids and thiol groups on proteins to form lipoperoxides, causing cell death. Both cell types create cytotoxicity in the following order: hGO > GO > rGO. Reactive oxygen species (ROS) produced by graphene-based materials can cause cell damage and mitochondrial dysfunction. Furthermore, ROS can promote lipid peroxidation in membrane lipids by forming lipid peroxide by interacting with unsaturated fatty acids.

Due to their small dimension and sharp edges, graphene-based materials can reach the cytoplasm, causing membrane breakdown and cytoplasm leakage. GO-based materials, according to these findings, can cause cell death in human HaCaT skin keratinocytes via ROS production, mitochondrial malfunction, and LDH leakage [5]. These materials also have the potential to interact directly with DNA, causing genotoxicity. Furthermore, the type of GO, exposure time, and dosage also influence the biocompatibility and toxicity of graphene-based nanomaterials.

The impact of pure graphene sheets on the biocompatibility and cytotoxicity of biological systems has shown varied results. At a high dosage of 10 µg/ml, graphene caused negative consequences by reducing mitochondrial activity and raising ROS, caspase-3, and LDH levels [6], according to Zang et al. Intercellular proteases that catalyze proteolytic cleavage are known as caspases. On the other side, Nayak et al. observed that CVD-grown graphene sheets do not negatively impact the proliferation of human mesenchymal stem cells (hMSCs) and, in fact, speed up their differentiation into bone cells [7]. Kim et al. also demonstrated that CVD-produced graphene sheets exhibit sustained cell attachment and are biocompatible with primary human cardiac cells [8]. Using LPCVD-produced graphene films, Rastogi et al. studied the survival and cell stress of both nonneuronal and neuronal cells [9]. According to the researchers, graphene enhances cell adhesion and proliferation in both cell types. Furthermore, neither cell type's MMP or morphology was affected by graphene, demonstrating that pure graphene does not produce cell stress. In vitro and in vivo osteogenic growth of hASCs and MSCs is stimulated by pristine graphene, according to Zhou et al. [10]. Chang et al. investigated how the size and dosage of GO exposed to A549 cells affected its biocompatibility [11]. With A549 cells, GO was determined to be biocompatible and did not induce cytotoxicity. It is worth noting that chemical agents used to reduce GO, such as anhydrous hydrazine, hydrazine monohydrate, and sodium borohydride, are extremely harmful to human cells. These should not be utilized in the synthesis of rGO as a reducing agent. Natural substances such as turquoise polyphenol, ascorbic acid, and glucose can be substituted. Dasgupta et al. successfully reduced GO in aqueous solution using polysaccharide extract from wild edible mushroom [12].

Understanding how GOs interact with the immune and circulatory systems is critical. GOs' target cells are macrophages [13]. When macrophages are activated, pristine graphene and GOs cause the production of several proinflammatory cytokines, including monocyte-derived cytokines like IL-1α, IL-6, IL-10, and tumor necrosis factor alpha (TNF-α), as well as chemokines, like monocyte to chemoattractant protein-1α (MIP-1α) and MIP-1β [14]. The ability of macrophages to internalize smaller GO was found to be stronger than the ability of macrophages to ingest larger GO. Because of their huge surface areas, large GOs are more likely to interact with cell membranes. According to Liao et al., GOs (765 ± 19 nm) induced hemolytic activity in human erythrocytes; however, sonicated GOs of a smaller size (342 ± 17nm) exhibited a greater hemolysis rate than large GOs [15]. On the other hand, coating GOs with chitosan almost eliminated hemolytic activity. The vitality of erythrocytes is thus influenced by the ways in which GOs interact with cells.

Wu et al. employed primary cells to investigate the toxicity of GOs [16]. These cells mimicked the physiological conditions of the cells *in vivo*. GOs did not cause cytotoxicity in human corneal epithelial cells after 2 h but generated significant cytotoxicity after 24 h, according to the data. The viability of cells decreases drastically as GO dosages increase. The WST-8 experiment demonstrates that when human conjunctiva epithelial cells (hConECs) are exposed to 12.5 µg/ml GO for 24 h, around 40% of hConECs died. They conclude that GOs produce dose- and time-dependent cytotoxicity in hCorECs and hConECs, as well as oxidative stress in response to ROS detection. The functionalization of GO has a big impact on its biocompatibility. Pegylated GOs show the highest biocompatibility with mice fibroblasts cell line when dosed at 3.125–12.5 µg/ml [17]. The relative cell viability for this dosage range is roughly 80%, but when the dose is raised to 50 µg/ml, cell viability drops to 60% and 40%, respectively. According to Luo et al., PEGylated GO sheets boosted immunological response in peritoneal macrophages by secreting cytokines and activating integrin beta-8-associated signaling pathways [18]. The survival of murine macrophages treated with GO, GO-NH$_2$, poly(acrylamide)-functionalized GO (GO-PAM), polyacrylic acid-functionalized GO (GO-PAA), and GO-PEG was examined using *in vitro* and *in vivo* studies by Xu et al. [19]. Upon the cytotoxicity investigations, GO-PEG and GO-PAA were determined to have the best biocompatibility. Table 9.1

enlists the observations reported regarding the biocompatibility and cytotoxicity assessment of different GOs and their functionalized derivatives.

9.2   BIOCOMPATIBILITY OF MXENE-BASED MATERIALS

There have been various investigations exploring the potential of MXene and MXene-based materials for different biomedical applications; however, the cytotoxicity and biocompatibility remains largely unexplored. MXenes' cytotoxicity appears to be dependent on its method of synthesis, oxidation state, mechanism of interaction, surface functional groups, size and route of administration, and exposure period [20]. Using MTT tests, Jastrzebka et al. investigated the *in vitro* cytotoxicity of delaminated $Ti_3C_2T_x$ MXene on two cancer cell lines (A549 and A375) and two normal cell lines (MRC-5 and HaCaT) [21]. Their findings revealed that $Ti_3C_2T_x$ MXene does not cause cytotoxicity at concentrations less than 62.5 mg/L. The cytotoxic mechanism of action on cancer cells, according to the article, is caused by the creation of ROS that exceeds the threshold, causing cancer cells to die. The research [22] revealed a substantial electrostatic interaction between the negatively charged MXene surface and the positively charged phosphatidylcholine lipids. The impact of MXene exposure to bone marrow–derived mesenchymal stem cells (BMSCs) has been investigated [23]. In addition to boosting BMSCs differentiation to osteoblasts, the MXene composite nanofibers showed good biocompatibility and increased cellular activity.

According to Scheibe et al., the *in vitro* cytotoxicity of various $Ti_3C_2T_z$ MXene and their precursors on human fibroblasts and HeLa cells is owing to the generation of oxidative stress in the cells upon exposure to high concentrations of TiC and MAX phases up to 400 µg/ml [24]. The biocompatibility of multilayered Ti_2NT_x MXene was investigated using MTT assays on normal human mammary epithelial cells (MCF-10A), malignant melanoma skin cells (A37), breast cancer cells (MCF-7), and immortalized keratinocytes (HaCaT) [25]. Ti_2N MXene like Ti_3C_2 showed greater toxicity to malignant cell lines as compared to normal cells. Jastrzebska et al. studied the cytotoxicity of single- and few-layered V2CTx MXene flakes using malignant melanoma (A375) and immortalized keratinocytes (HaCaT) human cell lines [20]. The reason of V2CTx cytotoxicity was revealed to be in situ oxidation of V2CTx flakes or vanadium oxide. The surface termination groups on $Ti_3C_2T_z$ were transformed to Ti_2O_3 via sonication and heat treatments [21]. According to the observations, thermally oxidized materials were selectively harmful to malignant cell lines up to 375 mg/L.

MXene toxicity has been connected to the generation of intracellular reactive oxygen species (ROS), which damage proteins and DNA, eventually leading to cell death. In the presence of water, MXene suspension dissociates into a radical hydroxide group ($\cdot OH$), a superoxide anion ($\cdot O_2^-$), and a hydrogen ion (H^+) [26]. Membrane breakdown was triggered by the peroxide anion generated by the radicals. Another reason is that MXene sheets make a strong contact with the cell membrane, causing cell instability and integrity loss. The size of the MXene sheet has significant impact on cell cytotoxicity. According to Jastrzebska et al. [21], MXene with nano lateral size can be endocytosed into cells. When the concentration was more than 25 µg/ml, TEM results revealed that neural stem cells were penetrated with the MXene of 2 µm width. MTT tests were used by Shi et al. in investigating the cytotoxicity of MXene flake/Cu NC nanoclusters in 3T3 murine fibroblasts and HEK293 cells using MTT assays [27]. Cell viability was determined to be 85% after treatment with MXene flakes and CuNCs.

In an *in vivo* mouse study, the animals were orally administered with 20 mg kg^{-1} of modified MXene, Nb_2C-PVP nanosheets [28]. In the tissue histological section or main organ staining, there were no unfavorable effects or pathological activities. The neurotoxicity and locomotion experiments revealed no significant effect even after 50 µg/L $Ti_3C_2T_x$ treatment. The *in vivo* cytotoxicity of $Ti_3C_2T_x$ MXene film implanted in subcutaneous and calvarial defect sites in rats was assessed by micro-CT and histological tests [29]. According to the findings, MXene films appeared to improve robust bone regeneration activity and osteo inductivity.

When MXene sheets are covered with biopolymers, cell cytotoxicity is greatly decreased. Non-malignant cells were shown to have high biocompatibility with Ti_2C treated with PEG, with vitality reaching 70% at all tested doses, compared to 50% before treatment [30]. Using osteoblasts, Liu et al. evaluated the cytotoxicity of PLGA-coated Ti_3C_2 MXene placed on Mg-Sr alloys [31]. According to cytotoxicity research utilizing the MTT test, magnesium coated with PLGA and $Ti_3C_2T_x$ had cell vitality of over 90% for 5 days, whereas magnesium had cell viability of over 80%. Few studies related to the biocompatibility of MXene are discussed in Table 9.1.

9.3 BIOCOMPATIBILITY OF TMDS-BASED MATERIALS

TMDs are gaining popularity in the biomedical field, although there have been few research articles reported exposing their biocompatibility and toxicity. Like graphene-based materials, the exfoliation process and lateral dimension of TMDs are important factors in determining their biocompatibility [32]. Teo et al. investigated the cytotoxicity of lithiated variants of WS_2, MoS_2, and WSe_2 in human lung cancer epithelial cells A549. WS_2 and MoS_2 were shown to be highly biocompatible, whereas lithiated WSe_2 caused dose-dependent toxicity [33]. Like other 2D materials, PEGylation of nanosheets or anchoring them with macromolecules like bovine serum albumin improves biocompatibility and cellular absorption. The PEGylated TMDs nanosheets of MoS_2, WS2, and TiS2 accumulated preferentially in the liver and spleen, although each of them has different residence times [34]. Despite the extended residence period, *in vitro* testing of PEGylated TMD nanosheets in mouse macrophage raw 264.7, human renal epithelial cell 293T, and mouse breast cancer 4T1 cell lines revealed no toxicity even at the maximum dosage. The addition of lipoic acid–modified PEG to MoS_2 sheets improved their physiological stability and biocompatibility [35]. The cytotoxicity of fullerenes, such as MoS_2 and WS_2, on human cell lines, salivary gland cells, and A549 cells revealed that they are harmless to cells, with cell viability remaining high following prolonged exposure to TMDs [36, 37].

TMDC toxicity studies on human lung carcinoma epithelial cells (A549) exposed to various concentrations of bulk TMDs (WSe_2, MoS_2, and WS_2) were carried out [33]. It was discovered that MoS_2 and WS_2 have no impact at greater concentrations up to 200 g/ml; however, WSe_2 demonstrated dose-dependent

TABLE 9.1

Biocompatibility Studies of Graphene-Based, TMDs-Based, and MXene-Based Materials

2D Nanomaterial	Cells	Cytotoxicity Effect	Reference
GO	RAW 264.7	Membrane pores apoptosis at ≥ 10 µg/mL	[2]
GO	A549	Toxicity dependent on dose; cell loss at ≥ 50 µg/mL	[2]
GO	Human skin keratinocyte	ROS generation, that is, dose- and time-dependent	[25]
GO	Human erythrocyte	Dose- and size-dependent hemolysis	[15]
GO-PEG	Saos-2; MC3T3-E1; RAW-264.7	GO-PEG accumulation on F-actin, resulting in production of ROS	[39]
GO-PAM	Murine macrophage	Cell death at ≥ 10 µg/mL	[19]
rGO-PEG	Murine astrocyte	Cell loss at >100 µg/mL	[40]
TRG	Monkey vero	Cell death at >100 µg/mL	[41]
$Ti_3C_2T_x$ and its precursors	Human fibroblasts and HeLa cells	Amount-dependent and cell-type-dependent cell loss	[24]
Ti_2C -PEG	A549 and A375	Cell viability up to 70% at 500 µg/mL in non-cancerous cells	[42]
$Ti_3C_2T_x$	A549 & A375	Cell toxicity depending on concentration	[25]
V_2CT_z MXene	Malignant melanoma (A375) and immortalized keratinocytes (HaCaT)	Dose- and time-dependent cell viability	[25]
Ti_3C_2 QDS	Human embryo kidney cells 2937 and MCF-7 cancer cells	Less cytotoxic	[43]
Ti_3C_2 and Nb_2C MXene quantum dots	Human umbilical vein endothelial cells (HUVECs)	Ti_3C_2 QDs could induce cytotoxicity to HUVECs at 100 µg/mL.	[27]
MXene flake/Cu NC	3 T3 mouse fibroblasts and HEK293 cells	No evidence of cytotoxicity	[44]
$Ti_3C_2T_x$ MXene films	Mouse pre-osteoblast cell line MC3T3E1	Enhanced osteogenic differentiation with high cytocompatibility	[29]
$Ti_3C_2T_z$/chitosan nanofibers	HeLa cells	Toxicity-free	[45]
Boron nitride (hBN)	Osteoblast-like cells ($SaOS_2$)	Unsaturated B atoms located on the particle surface causing cell death	[46]

TABLE 4.1 (CONTINUED)

2D Nanomaterial	Cells	Cytotoxicity Effect	Reference
MoS_2	Human acute monocytic leukemia (THP-1) cells, human lung adenocarcinoma (A549), and human gastric adenocarcinoma (AGS) cells	Exfoliated MoS_2 sheets non-toxic at values below 1 µg ml^{-1}	[47]
Atomically thin MoS_2 film and microparticles	Pancreatic cancer cell (PANC1), immortal kidney epithelial cell (293), pancreas cell (HPDE)	Cell viability dose-dependent; no discernible cytotoxic response on the film	[48]
MoS_2 and BN nanosheets	Human hepatoma HepG2 cells	At 30 µg/mL, both MoS_2 and BN significantly lowered cell viability	[49]
WS_2	A549 and human liver cancer cell line (HEP G2)	In A549, 50% reduction in cell viability compared to around 25% in the case of HEP G2	[50]
Chitosan functionalized MoS_2 (CS MoS_2) nanosheets	Human dermal fibroblasts (HDF)	Moderately lowered cell viability of HDF cells	[51]
Pristine MoS_2 and WS_2	Human epithelial kidney cells (HEK293f)	Toxicity-free	[52]

toxicity with a considerable loss in cell viability. The toxicity of WSe_2 has been linked to the cells being exposed to chalcogen rather than the transition metal W. In general, selenium and vanadium play important roles in toxicity, and ditellurides have greater cytotoxicity than materials containing disulfide [32]. On NIH/3T3 immortalized dermal fibroblasts cell line, the cytotoxicity of multilayered MoS_2 exfoliated using Li intercalation procedures and separately annealed to achieve 2H phase was investigated [38]. When compared to the non-toxic impact of the exfoliated MoS_2 sample without annealing, the annealed MoS_2 sample showed a substantially larger drop in cell viability. Some data related to TMDs are depicted in Table 9.1.

9.4 OTHER 2D MATERIALS

In addition to the aforementioned 2D materials, there are other layered 2D materials with atomic structures comparable to graphene, such as layer double hydroxides (LDHs), hexagonal boron nitride (hBN), and black phosphorous (BP). Like other 2D materials, the biocompatibility of hBN nanosheet is governed by its size, shape, form, and reactive surface. The hBN is biocompatible with osteoblast-like cells, but not with those with a diameter less than 1 µm and a depth less than 100 nm, according to the study [46]. The production of boron radicals on the nanosheets' surfaces produces cytotoxicity, which leads to cell death. In *in vivo* biocompatibility investigations, the planarians model did not show any cytotoxicity, although any coating may lessen the potential cytotoxicity [53].

In addition, BP has excellent biodegradability, outstanding biocompatibility, with few adverse effects, and tissue systemic toxicity. The 4T1 cell viability was more than 85% in the MTT experiment with PEGylated BP at a concentration less than 140 μg/ml [54]. In addition, the *in vivo* investigation revealed that BP had no substantial toxicity to major organs, such as liver, heart, lung, spleen, and kidneys. Furthermore, the diameter of BP nanosheets showed no toxic effect on the biocompatibility parameters [55].

The biocompatibility of LDH was determined using MTT assay, which revealed that after 24 h of incubation with various concentrated LDH-Gd/Au medium solutions, the survival rate of L929 cells and HeLa cells remained greater than 90% [56]. The *in vivo* histological assessment against liver, spleen, heart, lung, kidney, and spleen of KM mice treated with LDH-Gd/Au showed no evident histopathological testing, showing good *in vivo* biocompatibility with the composite. Similarly, other 2D materials, like 2D clay nanosheets, metal organic framework, and covalent organic nanosheets, have also aroused interest in the biomedical field, for which in-depth biocompatibility study is needed for their better application.

REFERENCES

1. Donaruma LG. Definitions in biomaterials. In: D.F. Williams, Ed. Amsterdam: Elsevier; 1987:414.
2. Duan G, Zhang Y, Luan B, Weber JK, Zhou RW, Yang Z, Zhao L, Xu J, Luo J, Zhou R. Graphene-induced pore formation on cell membranes. Scientific Reports. 2017;7:42767.
3. Frost R, Jönsson GE, Chakarov D, Svedhem S, Kasemo B. Graphene oxide and lipid membranes: interactions and nanocomposite structures. Nano Letters. 2012;12:3356–62.
4. Li R, Guiney LM, Chang CH, Mansukhani ND, Ji Z, Wang X, Liao Y-P, Jiang W, Sun B, Hersam MC, Nel AE, Xia T. Surface oxidation of graphene oxide determines membrane damage, lipid peroxidation, and cytotoxicity in macrophages in a pulmonary toxicity model. ACS Nano. 2018;12:1390–402.
5. Pelin M, Fusco L, Martín C, Sosa S, Frontiñán-Rubio J, González-Domínguez JM, Durán-Prado M, Vázquez E, Prato M, Tubaro A. Graphene and graphene oxide induce ROS production in human HaCaT skin keratinocytes: the role of xanthine oxidase and NADH dehydrogenase. Nanoscale. 2018;10:11820–30.
6. Zhang Y, Ali SF, Dervishi E, Xu Y, Li Z, Casciano D, Biris AS. Cytotoxicity effects of graphene and single-wall carbon nanotubes in neural phaeochromocytoma-derived PC12 cells. ACS Nano. 2010;4:3181–6.
7. Nayak TR, Andersen H, Makam VS, Khaw C, Bae S, Xu X, Ee P-LR, Ahn J-H, Hong BH, Pastorin G, Özyilmaz B. Graphene for controlled and accelerated osteogenic differentiation of human mesenchymal stem cells. ACS Nano. 2011;5:4670–8.
8. Kim T, Kahng YH, Lee T, Lee K, Kim DH. Graphene films show stable cell attachment and biocompatibility with electrogenic primary cardiac cells. Molecules and cells. 2013;36:577–82.
9. Rastogi SK, Raghavan G, Yang G, Cohen-Karni T. Effect of graphene on nonneuronal and neuronal cell viability and stress. Nano Letters. 2017;17:3297–301.
10. Liu Y, Chen T, Du F, Gu M, Zhang P, Zhang X, Liu J, Lv L, Xiong C, Zhou Y. Single-layer graphene enhances the osteogenic differentiation of human mesenchymal stem cells in vitro and in vivo. Journal of Biomedical Nanotechnology. 2016;12:1270–84.

11. Chang Y, Yang ST, Liu JH, Dong E, Wang Y, Cao A, Liu Y, Wang H. In vitro toxicity evaluation of graphene oxide on A549 cells. Toxicology Letters. 2011;200:201–10.
12. Dasgupta A, Sarkar J, Ghosh M, Bhattacharya A, Mukherjee A, Chattopadhyay D, Acharya K. Green conversion of graphene oxide to graphene nanosheets and its biosafety study. PLoS One. 2017;12:e0171607.
13. Zhang B, Wei P, Zhou Z, Wei T. Interactions of graphene with mammalian cells: molecular mechanisms and biomedical insights. Advanced Drug Delivery Reviews. 2016;105:145–62.
14. Russier J, Treossi E, Scarsi A, Perrozzi F, Dumortier H, Ottaviano L, Meneghetti M, Palermo V, Bianco A. Evidencing the mask effect of graphene oxide: a comparative study on primary human and murine phagocytic cells. Nanoscale. 2013;5:11234–47.
15. Liao K-H, Lin Y-S, Macosko CW, Haynes CL. Cytotoxicity of graphene oxide and graphene in human erythrocytes and skin fibroblasts. ACS Applied Materials & Interfaces. 2011;3:2607–15.
16. Wu W, Yan L, Wu Q, Li Y, Li Q, Chen S, Yang Y, Gu Z, Xu H, Yin ZQ. Evaluation of the toxicity of graphene oxide exposure to the eye. Nanotoxicology. 2016;10:1329–40.
17. Du L, Wu S, Li Y, Zhao X, Ju X, Wang Y. Cytotoxicity of PEGylated graphene oxide on lymphoma cells. Bio-medical Materials and Engineering. 2014;24:2135–41.
18. Luo N, Weber JK, Wang S, Luan B, Yue H, Xi X, Du J, Yang Z, Wei W, Zhou R, Ma G. PEGylated graphene oxide elicits strong immunological responses despite surface passivation. Nature Communications. 2017;8:14537.
19. Xu M, Zhu J, Wang F, Xiong Y, Wu Y, Wang Q, Weng J, Zhang Z, Chen W, Liu S. Improved in vitro and in vivo biocompatibility of graphene oxide through surface modification: poly(acrylic acid)-functionalization is superior to PEGylation. ACS Nano. 2016;10:3267–81.
20. Jastrzębska AM, Scheibe B, Szuplewska A, Rozmysłowska-Wojciechowska A, Chudy M, Aparicio C, Scheibe M, Janica I, Ciesielski A, Otyepka M, Barsoum MW. On the rapid in situ oxidation of two-dimensional V(2)CT(z) Mxene in culture cell media and their cytotoxicity. Materials Science & Engineering C, Materials for Biological Applications. 2021;119:111431.
21. Jastrzębska A, Szuplewska A, Wojciechowska A, Chudy M, Olszyna A, Birowska M, Popielski M, Majewski J, Scheibe B, Natu V, Barsoum M. On tuning the cytotoxicity of Ti3C2 (Mxene) flakes to cancerous and benign cells by post-delamination surface modifications. 2D Materials. 2020;7.
22. Ou L, Song B, Liang H, Jia L, Feng X, Deng B, Sun T, Shao L. Toxicity of graphene-family nanoparticles: a general review of the origins and mechanisms. Particle and Fibre Toxicology. 2016;13.
23. Huang R, Chen X, Dong Y, Zhang X, Wei Y, Yang Z, Li W, Guo Y, Liu J, Yang Z, Wang H, Jin L. Mxene composite nanofibers for cell culture and tissue engineering. ACS Applied Bio Materials. 2020;3:2125–31.
24. Scheibe B, Wychowaniec JK, Scheibe M, Peplińska B, Jarek M, Nowaczyk G, Przysiecka Ł. Cytotoxicity assessment of Ti-Al-C based MAX phases and Ti_3C_2Tx Mxenes on human fibroblasts and cervical cancer cells. ACS Biomaterials Science & Engineering. 2019;5:6557–69.
25. Jastrzębska AM, Szuplewska A, Wojciechowski T, Chudy M, Ziemkowska W, Chlubny L, Rozmysłowska A, Olszyna A. In vitro studies on cytotoxicity of delaminated Ti_3C_2 Mxene. Journal of Hazardous Materials. 2017;339:1–8.
26. Ganguly P, Breen A, Pillai SC. Toxicity of nanomaterials: exposure, pathways, assessment, and recent advances. ACS Biomaterials Science & Engineering. 2018;4:2237–75.
27. Shi Y-e, Han F, Xie L, Zhang C, Li T, Wang H, Lai W-F, Luo S, Wei W, Wang Z, Huang Y. A Mxene of type Ti_3C_2Tx functionalized with copper nanoclusters for the fluorometric determination of glutathione. Microchimica Acta. 2019;187:38.

28. Liu G, Zou J, Tang Q, Yang X, Zhang Y, Zhang Q, Huang W, Chen P, Shao J, Dong X. Surface modified Ti_3C_2 Mxene nanosheets for tumor targeting photothermal/photodynamic/chemo synergistic therapy. ACS Applied Materials & Interfaces. 2017;9:40077–86.

29. Zhang J, Fu Y, Mo A. Multilayered titanium carbide Mxene film for guided bone regeneration. International Journal of Nanomedicine. 2019;14:10091–103.

30. Szuplewska A, Rozmysłowska-Wojciechowska A, Poźniak S, Wojciechowski T, Birowska M, Popielski M, Chudy M, Ziemkowska W, Chlubny L, Moszczyńska D, Olszyna A, Majewski JA, Jastrzębska AM. Multilayered stable 2D nano-sheets of Ti2NTx Mxene: synthesis, characterization, and anticancer activity. Journal of Nanobiotechnology. 2019;17:114.

31. Liu L, Huang B, Liu X, Yuan W, Zheng Y, Li Z, Yeung KWK, Zhu S, Liang Y, Cui Z, Wu S. Photo-controlled degradation of PLGA/Ti_3C_2 hybrid coating on Mg-Sr alloy using near infrared light. Bioactive Materials. 2021;6:568–78.

32. Guiney LM, Wang X, Xia T, Nel AE, Hersam MC. Assessing and mitigating the hazard potential of two-dimensional materials. ACS Nano. 2018;12:6360–77.

33. Teo WZ, Chng ELK, Sofer Z, Pumera M. Cytotoxicity of exfoliated transition-metal dichalcogenides (MoS_2, WS_2, and Wse_2) is lower than that of graphene and its analogues. Chemistry. 2014;20:9627–32.

34. Hao J, Song G, Liu T, Yi X, Yang K, Cheng L, Liu Z. In vivo long-term biodistribution, excretion, and toxicology of PEGylated transition-metal dichalcogenides MS_2 (M = Mo, W, Ti) nanosheets. Advanced Science. 2017;4:1600160.

35. Liu T, Wang C, Gu X, Gong H, Cheng L, Shi X, Feng L, Sun B, Liu Z. Drug delivery with PEGylated MoS_2 nano-sheets for combined photothermal and chemotherapy of cancer. Advanced Materials. 2014;26:3433–40.

36. Goldman EB, Zak A, Tenne R, Kartvelishvily E, Levin-Zaidman S, Neumann Y, Stiubea-Cohen R, Palmon A, Hovav A-H, Aframian DJ. Biocompatibility of tungsten disulfide inorganic nanotubes and fullerene-like nanoparticles with salivary gland cells. Tissue Engineering Part A. 2015;21:1013–23.

37. Wu H, Yang R, Song B, Han Q, Li J, Zhang Y, Fang Y, Tenne R, Wang C. Biocompatible inorganic fullerene-like molybdenum disulfide nanoparticles produced by pulsed laser ablation in water. ACS Nano. 2011;5:1276–81.

38. Fan J, Li Y, Nguyen HN, Yao Y, Rodrigues DF. Toxicity of exfoliated-MoS_2 and annealed exfoliated-MoS_2 towards planktonic cells, biofilms, and mammalian cells in the presence of electron donor. Environmental Science: Nano. 2015;2:370–9.

39. Matesanz MC, Vila M, Feito MJ, Linares J, Gonçalves G, Vallet-Regi M, Marques PA, Portolés MT. The effects of graphene oxide nanosheets localized on F-actin filaments on cell-cycle alterations. Biomaterials. 2013;34:1562–9.

40. Mendonça MC, Soares ES, de Jesus MB, Ceragioli HJ, Batista Â G, Nyúl-Tóth Á, Molnár J, Wilhelm I, Maróstica MR, Jr., Krizbai I, da Cruz-Höfling MA. PEGylation of reduced graphene oxide induces toxicity in cells of the blood-brain barrier: an in vitro and in vivo study. Molecular Pharmaceutics. 2016;13:3913–24.

41. Sasidharan A, Panchakarla LS, Chandran P, Menon D, Nair S, Rao CNR, Koyakutty M. Differential nano-bio interactions and toxicity effects of pristine versus functionalized graphene. Nanoscale. 2011;3:2461–4.

42. Szuplewska A, Kulpińska D, Dybko A, Jastrzębska AM, Wojciechowski T, Rozmysłowska A, Chudy M, Grabowska-Jadach I, Ziemkowska W, Brzózka Z, Olszyna A. 2D Ti_2C (Mxene) as a novel highly efficient and selective agent for photothermal therapy. Materials Science and Engineering: C. 2019;98:874–86.

43. Zhou L, Wu F, Yu J, Deng Q, Zhang F, Wang G. Titanium carbide (Ti_3C_2Tx) Mxene: a novel precursor to amphiphilic carbide-derived graphene quantum dots for fluorescent ink, light-emitting composite and bioimaging. Carbon. 2017;118:50–7.

44. Gu M, Dai Z, Yan X, Ma J, Niu Y, Lan W, Wang X, Xu Q. Comparison of toxicity of Ti_3C_2 and Nb_2C Mxene quantum dots (QDs) to human umbilical vein endothelial cells. 2021;41:745–54.

45. Mayerberger EA, Street RM, McDaniel RM, Barsoum MW, Schauer CL. Antibacterial properties of electrospun Ti_3C_2Tz (Mxene)/chitosan nanofibers. RSC Advances. 2018;8:35386–94.

46. Mateti S, Wong CS, Liu Z, Yang W, Li Y, Li LH, Chen Y. Biocompatibility of boron nitride nanosheets. Nano Research. 2018;11:334–42.

47. Liu C, Kong D, Hsu P-C, Yuan H, Lee H-W, Liu Y, Wang H, Wang S, Yan K, Lin D, Maraccini PA, Parker KM, Boehm AB, Cui Y. Rapid water disinfection using vertically aligned MoS_2 nanofilms and visible light. Nature Nanotechnology. 2016;11:1098–104.

48. Murugan C, Murugan N, Sundramoorthy AK, Sundaramurthy A. Nanoceria decorated flower-like molybdenum sulphide nanoflakes: an efficient nanozyme for tumour selective ROS generation and photo thermal therapy. Chemical Communications. 2019;55:8017–20.

49. Liu S, Shen Z, Wu B, Yu Y, Hou H, Zhang X-X, Ren H-Q. Cytotoxicity and efflux pump inhibition induced by molybdenum disulfide and boron nitride nanomaterials with sheet-like structure. Environmental Science & Technology. 2017;51:10834–42.

50. Liu X, Duan G, Li W, Zhou Z, Zhou R. Membrane destruction-mediated antibacterial activity of tungsten disulfide (WS2). RSC Advances. 2017;7:37873–80.

51. Yu Y, Wu N, Yi Y, Li Y, Zhang L, Yang Q, Miao W, Ding X, Jiang L, Huang H. Dispersible MoS_2 nanosheets activated TGF-β/smad pathway and perturbed the metabolome of human dermal fibroblasts. ACS Biomaterials Science & Engineering. 2017;3:3261–72.

52. Appel JH, Li DO, Podlevsky JD, Debnath A, Green AA, Wang QH, Chae J. Low Cytotoxicity and genotoxicity of two-dimensional MoS_2 and WS_2. ACS Biomaterials Science & Engineering. 2016;2:361–7.

53. Ciofani G, Danti S, Nitti S, Mazzolai B, Mattoli V, Giorgi M. Biocompatibility of boron nitride nanotubes: an up-date of in vivo toxicological investigation. International Journal of Pharmaceutics. 2013;444:85–8.

54. Shao J, Xie H, Huang H, Li Z, Sun Z, Xu Y, Xiao Q, Yu X-F, Zhao Y, Zhang H, Wang H, Chu PK. Biodegradable black phosphorus-based nanospheres for in vivo photothermal cancer therapy. Nature Communications. 2016;7:12967.

55. Fu H, Li Z, Xie H, Sun Z, Wang B, Huang H, Han G, Wang H, Chu PK, Yu X-F. Different-sized black phosphorus nanosheets with good cytocompatibility and high photothermal performance. RSC Advances. 2017;7:14618–24.

56. Wang L, Xing H, Zhang S, Ren Q, Pan L, Zhang K, Bu W, Zheng X, Zhou L, Peng W, Hua Y, Shi J. A Gd-doped Mg-Al-LDH/Au nanocomposite for CT/MR bimodal imagings and simultaneous drug delivery. Biomaterials. 2013;34:3390–401.

10 Conclusions and Future Prospects

Neeraj Dwivedi and Avanish Kumar Srivastava

In this book, we have comprehensively discussed the synthesis, fundamental properties, and biomedical applications – in particular, in the context of COVID-19 – of emerging 2D materials, such as graphene, GO, rGO, MoS_2, MAX, MXene, and so on. Firstly, the structure of coronavirus, including SARS-CoV-2, was discussed in detail. Subsequently, various physical and chemicals methods were discussed for the synthesis of these 2D materials. These include micromechanical cleavage, liquid-phase exfoliation, ion-intercalated exfoliation, wet-chemical synthesis, chemical vapor deposition methods, and so on. Additionally, the physicochemical and functional properties of these materials have also been discussed. Antiviral characteristic is essential to combat COVID-19 through materials approach. Therefore, we comprehensively and critically discussed the antiviral efficacy of graphene-related materials, TMDCs, and MXene, and MAX materials. It was found that graphene, GO, rGO, MoS_2, MXene, their hybrids and composites display excellent antiviral properties, and they inactivate/kill viruses by a number of mechanisms, including rupturing *via* sharp edges of the materials, oxidative stress, etc. Further, we also discussed the progress in the development of a variety of biosensors, including electrochemical, FET, and piezo-based biosensors, for early detection of pathogens, including coronavirus. Additionally, underlying mechanisms for the detection of pathogens, sensitivity, and accuracy of biosensors based on 2D materials were also presented. Advanced manufacturing processes can also contribute to efficiently combating the pandemic. Additive manufacturing is one such advanced manufacturing process that can even print the complex component designs. Additive manufacturing has been widely employed during the COVID-19 pandemic to develop various PPEs, medical components, testing devices, etc., which all were thoroughly discussed in the book. Further, the application of foam materials, including the foams based on 2D materials for the development of components for COVID-19 and other pandemic, was also comprehensively discussed in the book. The sterilizer is a very effective device for the inactivation of viruses and other microorganisms. We have comprehensively discussed the photothermal and photocatalytic properties of graphene, GO, rGO, MoS_2, MXene, and their hybrids and proposed how these materials can be exploited to develop next-generation sterilizers. Finally, the biocompatibility and cytotoxicity characteristics of diverse 2D materials have been discussed.

While 2D materials have shown remarkable biomedical properties at lab scale, the devices or the systems based on any kind of 2D materials, whether MXene, TMDCs, or even graphene, have not yet progressed to clinical trial stage. Healthcare

DOI: 10.1201/9781003316381-10

products based on these 2D materials have yet to be launched in the market. So, the path for 2D materials–based medical systems is far long for commercialization. Thus, one of the most essential steps in 2D materials research is to extensively perform R&D at clinical trial stage. This includes the development and testing of 2D materials–based antiviral components/systems, PPEs, sterilizers, biosensors, etc. In particular for COVID-19, various medical components based on 2D materials must be tested against SARS-CoV-2 at clinical trial or equivalent stages as well. Further, MXene-based materials have also shown exceptional antimicrobial properties, which were found to exceed even those of graphene or GRMs. However, for practical applications, for example, in hospital settings, how MXene-based coatings can be applied on large-scale components, such as glass windows or glass panels, is yet to be explored. Thus, development and demonstration of large-area antimicrobial/antiviral coatings based on 2D materials are another promising subject of 2D materials research for biomedical applications. Next, in order to explore materials for biomedical applications, biocompatibility and cytotoxicity are the critical factors. Thus, before deploying 2D materials–based components on the field, their biocompatibility and cytotoxicity must be investigated extensively. In the end, while the 2D materials–based commercial medical devices are not available in the market so far, however, due to their exceptional antiviral, biosensing, photocatalytic, photothermal, and other functional characteristics, the future of 2D materials for biomedical application is bright. This indicates that in the future, commercial biomedical components can employ 2D materials during their production.

Index